deep diversity

shakil choudhury

deep diversity

a compassionate, scientific approach to achieving racial justice

GREYSTONE BOOKS
Vancouver/Berkeley

Published by Greystone Books in 2021
First edition published as *Deep Diversity: Overcoming Us vs. Them* in 2015 by Between the Lines

21 22 23 24 25 5 4 3 2 1

Some names have been changed to protect the identity of those involved. Every effort has been made to secure permission for all material quoted or recounted.

Greystone Books Ltd.
greystonebooks.com

Cataloging data available from Library and Archives Canada
ISBN 978-1-77164-901-8 (cloth)
ISBN 978-1-77164-902-5 (epub)

Editing on first edition by Mary Newberry
Editing on second edition by Jennifer Croll
Copyediting by Jess Schulman
Proofreading by Meg Yamamoto
Indexing by Alison Strobel
Jacket and text design by Jessica Sullivan
Author photo by Darius Bashar

Printed in Canada on FSC® certified paper at Friesens. The FSC® label means that materials used for the product have been responsibly sourced.

Greystone Books gratefully acknowledges the Musqueam, Squamish, and Tsleil-Waututh peoples on whose land our office is located.

Greystone Books thanks the Canada Council for the Arts, the British Columbia Arts Council, the Province of British Columbia through the Book Publishing Tax Credit, and the Government of Canada for supporting our publishing activities.

Canada

To Arion, Koda, and the
next generation...
May you
Build on our successes and
learn from our mistakes,
Grow the Beloved Community
to shelter from the storm,
Use the sacred fire of hope
to feed your courage.

Contents

Preface

DEEP DIVERSITY is an approach to identifying, understanding, and tackling the hidden and unconscious ways in which racism shows up in society today, including within ourselves. Deep Diversity is unusual as it blends the principles of compassion, justice, and psychology, like imagining a kitchen-table conversation where the Dalai Lama, Black Lives Matter, and Carl Jung find common ground.

Existing racial tensions and social divisions have only increased as a result of COVID-19, Donald Trump, and the George Floyd protests with global repercussions. We are in a post-pandemic world that has brought our health and international economy to its knees. Described by many as the most severe crisis since World War Two, the devastation includes a global death toll in the millions since the first cases of COVID-19 were detected[1] at the end of 2019. It's remarkable to remember that early in the crisis, half the world's population—over three billion people—was in extended lockdown or quarantine *at the same time*—a uniquely singular moment in history. The damage to national economies has been extensive with major industries hard hit including tourism, airlines, travel, movie/TV production, sports, and arts, not to mention the millions of small businesses that struggled to survive or that shut down entirely.

But the coronavirus served to exacerbate problems that were already present.

It came on the coattails of an era politically defined by the emergence of right-wing populist leaders with nationalistic, protectionist, and authoritarian tendencies. The most successful and powerful was the ascendancy of Donald Trump to the United States presidency in 2016. Trump changed the rules of the game and altered all expectations of political leadership with his brash, media-savvy, take-no-prisoners style of communication and unpredictable decision-making. His Twitter tirades simultaneously delighted his base and enraged his opponents.

Although Trump lost the 2020 election to Joe Biden, the brazen mob attack by his supporters on the Capitol Building in Washington on January 6, 2021, as well as firm allegiance by almost all Republican leaders who continued to defend him publicly even at the time of this book's completion, indicate we have not entered a Trump-free world.

It is safe to say that Trump divided the electorate into Us/Them like no prior U.S. president, damaging the social fabric of American democracy in the process. The rise of white nationalism and hate crimes, "fake news," Russian interference, corruption, and impeachment were just a few of the polarizing hallmarks of his four years in office. More ominous were the country's lurching leaps toward authoritarianism with Trump's contempt for the rule of law and democratic traditions, continual attacks on journalists, and open admiration of current-day dictators like Russia's Vladimir Putin and Kim Jong-Un of North Korea. Even following his defeat in the 2020 general election, he refused to concede to Joe Biden, aggressively spreading conspiracy theories that the election was "rigged," in spite of the complete lack of evidence. With widespread documentation of his pathological lying, Trump greatly eroded the key ingredient for democracy—*trust* in social institutions and between common citizens, a wound that may take generations to heal.

Although we have entered a post-Trump era, the legacy of Trumpism has had many global impacts including motivating extremists and nationalists in many countries such as Canada, the United Kingdom, and Germany. Trump's manner and attitudes were also copycatted by authoritarian leaders around the world. For example, his common accusation that any media critical of his words, view, or agenda was "fake news" was used by Poland's Andrzej Duda in order to sign a bill giving his government broad control over the media, Rodrigo Duterte in the Philippines to cover up the murder of thousands of people by his forces, and Bashar al-Assad of Syria to deny responsibility for thousands of secret deaths of political opponents in prison.[2]

Race/ethnicity has been a consistent through line in the story not only because Trump's base is disproportionately white, but because of his regular race-baiting tactics and instincts for the politics of division. He consistently stereotyped, disparaged, and scapegoated many minoritized groups including African Americans, Mexicans, Muslims, Jews, women, people with disabilities, and people of Chinese ethnicity. Furthermore, the COVID-19 crisis often hit the poorest communities hardest with Black, Indigenous, and people of color facing disproportionately high rates of sickness and death due to a variety of socioeconomic factors including overrepresentation in low-paying, low-skilled jobs in places like meatpacking plants and factories that made physical distancing impossible, and overcrowded housing or inadequate health care services.[3] As it was often said, *the virus does not discriminate, but society does.*

Further fuel to the fire was the brutal murder of George Floyd at the hands of a white police officer caught on video, which ignited massive protests and despair-fueled racialized violence not seen in the United States since the assassination of Martin Luther King Jr. in 1968. Over 150 cities were the sites of massive civilian uprisings, spiraling into a global movement with solidarity protests against anti-Black racism and police brutality spanning from London and Paris to Frankfurt, Sydney, and Tokyo.

The public continues to be deeply divided along partisan lines today and distrusting not only of government and politicians but of each other. Extreme polarization has become the unfortunate norm in the U.S. and in many Western nations with liberals and conservatives seeing each other as a threat to their nations' very existence. Race and identity have been, and continue to be, defining themes.

The level of polarization needs *immediate* course correction not only because it is incredibly unproductive, but because it is how democracies die, according to researchers of authoritarianism.[4] The time period we are in—and how we respond to it—is best named as the *bridging* or *breaking* moment, explained by legal scholar and African American elder john powell from the Othering & Belonging Institute at University of California, Berkeley.[5] We will either build common ground or we fall prey to the politics of division and isolation. The former allows us to thrive and strengthen collectively while the latter is the path of mutual destruction. Quite literally, democracy itself is on the line. The conditions are fertile for authoritarianism and violence to take root in industrialized societies, and especially in the United States, even in a post-Trump era.

Deep Diversity was first published in Canada in 2015, and the original book was designed to address the post-9/11 context, one that was already significantly alive with Us versus Them dynamics. Whatever the challenges were then have only become more extreme, with nationalism rising and identities hardening. This updated edition of *Deep Diversity* has been revised and expanded with new insights and research to contend with our current, even more fractured reality. It is also a call to action for bridge-builders who are needed now more than ever to step into the divide and nurture understanding and heal relationships. To help slow down—or, better yet, reverse—the Us/Them dynamic, it's critical to understand one of its foundational elements: *systemic racial discrimination*, also known as *racism*, which is the focus of this book.

To achieve racial justice, we need to first understand the problem we face in depth, to see how the historical patterns of race and identity have resulted in unequal access to justice and basic resources such as housing, health care, education, and employment for minoritized communities today. *Deep Diversity* offers a proven method for *seeing* the world with a clear yet compassionate perspective in order to *act* in ways that help dismantle racial injustice in both our personal and professional lives.

My Story: From Activism to Burnout... and Back Again

Deep Diversity seeks to reframe the debate regarding racism and systemic discrimination in a practical, scientific, and compassionate manner. It is intimately tied to my personal and professional story, a culmination of twenty-five years of experience in the field of racial justice, one emotional burnout at age thirty, and a childhood pretending I was white.

This origin story first requires a quick detour through my professional life. I'm the cofounder of Anima Leadership, a boutique firm that helps leaders nurture workplace cultures that are more inclusive, diverse, and equitable. We do this through a variety of ways including serving as strategic advisors for executive teams, offering training to managers, and conducting organizational audits, focus groups, and demographic surveys.

I've had the honor of teaching, working with, and learning from thousands of leaders and hundreds of organizations in North America, Central and South America, Europe, and South Asia. That's lots of grist for the mill on the topics of Us/Them, race, identity, and culture, including projects that were substantial successes, fabulous failures, and everything in between.

There are four basic approaches in the work I do: *multiculturalism*, *cross-cultural communications*, the *business case for diversity*, and *anti-racism and anti-oppression* (ARAO). Each of these comes with its strengths and weaknesses.[6] The first three are easier to engage yet can candy-coat or substantively avoid addressing issues of systemic discrimination altogether. ARAO, on the other hand, clearly identifies the problems of injustice and oppression yet can activate feelings of shame and blame, with or without intention.

Regardless of the differences between the approaches, I see these as strategies created to help both individuals and groups nurture environments in which all people feel they matter and belong, with desired outcomes related to fairness and justice. Boiled down to its essence, this work strives to increase the sense of "Us" while reducing the feelings of "Them."

Of the approaches listed, the last one mentioned, anti-racism/anti-oppression, is the most challenging and contentious of the bunch. It's also my background, as my master's thesis was focused on ARAO and was the foundation for my work early in my career.

Why ARAO is not as dominant today in my work and life as it once was is a key part of the backstory of Deep Diversity, of why it came to be. Let me explain.

Anti-racism is a political theory founded on the premise that racism can be eliminated, but to do so, power and its abuses must be addressed on both individual and institutional levels. For this to be meaningful, significant emphasis must be placed on the change being system-wide. As the theory goes, in a society where racism exists, it is not enough to be nonracist; one has to actively challenge discrimination in all its forms for real transformation to occur.

Thus, the term "anti-racism" is used, and "anti-oppression" more broadly to discuss how other forms of marginalization such as sexism, heterosexism/homophobia, or ableism also interlock and connect.

The broad goal of ARAO is to create a barrier-free society in which all people have the right to freedom and dignity, with access to resources and opportunities to help individuals and communities achieve their human potential. ARAO also recognizes that all such issues are *political*, that there is no such thing as being *non-political*—being neutral only serves to replicate a system that is unjust in the first place, and maintains oppression.

For a long time, I exclusively used this particular worldview as it helped me make sense of my life as a racial minority growing up in a white nation and struggling with feelings of inferiority. Although I was a popular, high-achieving kid, I grew up trying to hide my South Asian heritage, an experience of shame that I would later learn is fairly common among many minority group members. I went as far as completely avoiding other brown kids in my desperate bid to fit into a white society, frequently behaving as if, and believing that I was, white.

At the time, I didn't have the language to recognize—let alone describe—the feelings of inadequacy with which I wrestled. It wasn't until doing my master's degree in my mid-twenties that I started to make sense of my experience using an anti-racism lens and, more broadly, anti-oppression principles. I encountered government reports, commissions, and umpteen studies that clearly demonstrated how Black, Indigenous, and people of color (BIPOC) were treated worse than their white counterparts on so many social measures, such as access to jobs, pay, health care, education, or fairness in the justice system throughout North America. There was even a term to describe my childhood feelings of cultural shame and rejection: "internalized racism." The quantity and clarity of information was both overwhelming and empowering. The patterns of racial injustice became crystal clear; this was something that once seen could not be unseen.

I felt shocked, furious—and altered—by what I was learning. I was stunned that I had never been taught about systemic discrimination before. The theory meshed with my experiences, helping make sense of much of my life. I felt very powerful, in control in a personal way that I hadn't previously. Although I didn't know it at the time, this would become my life's work. Already having been trained as a teacher professionally, my identity now transformed into that of an anti-racist educator and activist, with all the trappings that come with the identity including a newfound sense of purpose and authority, not to mention a touch of born-again fervor.

By the time I was thirty, I'd accomplished a number of things. Some of the highlights included managing community projects in Costa Rican rain forests, coordinating an oral-history project between young leaders in Pakistan and Canada, and spearheading economic/political literacy workshops for low-income communities locally. I was a founding teacher of an alternative school in Toronto and had put in countless volunteer hours for community-based organizations around the city. I was also getting accolades for my work as an anti-racist educator, having been honored by a provincial award for anti-bias curriculum development.

But things were not all rosy in Activist Land.

There was a downside to being well versed in anti-racism content and learning processes: I perceived racism, discrimination, and oppression everywhere I went. This became an unconscious habit, an exclusive view on life (which was a little gloomy, to say the least).

We lived in a bleak, unjust planet that was sinking fast. Saving the world was a thankless, never-ending task that also seemed without choice—I felt compelled to do it (and resentful that most of society seemed to be unaware or not care). I prioritized my work and the needs of strangers ahead of time with my loved ones, who began to wonder why I wasn't around and why I was always so exhausted.

I was entering the territory of personal burnout and didn't know it.

There was also an emerging realization that the social justice community I admired and was inspired by had its own share of power dynamics, toxicity, and egos. In spite of the philosophies, ideologies, and mandates of many progressive organizations, we were hardly models of healthy relationships, making our criticism of mainstream or corporate organizations feel hollow. Our relationships were just as fractured as anyone else's, so our belief systems had little hope of being lived out in their fullness.

I recall a specific situation that unmoored me shortly after the 9/11 terrorist attacks in the United States in 2001. In a meeting, I watched a group of my activist peers bicker and snipe at one another as they tried to decide how to respond to a tragedy of such immense magnitude. Various worldviews were competing to influence the room including, but not limited to, anti-war, anti-racism, anti-globalization, anti-poverty, union, direct action, and feminist perspectives. The environment was sharply divided, political, and terribly unfriendly—surprising, considering that these people supposedly were all working toward a socially just world. Feelings of "Us" and "Them" were in the room but we couldn't see it.

Anger and frustration boiled over inside me.

This is my community? This is who I look up to and whose affirmation I desire? If we can't keep it together, who can? I don't need this—I'm out!

I was overstretched as it was. Instead of engaging in the room, I began to detach.

This detachment also occurred in my personal life, which was unbalanced with many relationships already fraying. I was unable to meet my obligations to those I loved the most. I felt resentful about giving so much of my time to the outside world and began to question what I was doing and why. I was worn out, emotionally adrift in grayness.

I walked away from community organizing and activism.

In retrospect, burning out was the best thing that could have happened to me. I was forced to embark on a painful but important journey of healing and self-discovery. Over the next decade, I leveraged my middle-class privilege and networks to find supports through friends, mentors, coaches, therapists, and trusted colleagues, uncovering unhealthy inter- and intrapersonal patterns, acknowledging old wounds and being more intentional about my life.

I began to realize that the common element for dysfunction in my life mostly had to do with me—my choices, actions, and reactions. I began to understand that infuriating yet liberating lesson plainly stated by wise elders such as the Dalai Lama and Epictetus: although we rarely control our circumstances, we always have *choice* over how to react to them.

The tricky part was starting to uncover the *unconscious* aspects of what I thought, said, and did that caused me trouble. Over a process of years, I became more aware, developing better habits in the choices I made and how I managed myself, especially in stressful and charged situations. I felt more in control as well as more spacious. The feelings of being tossed about by the waves of existence with limited personal choice began to recede. I was also learning an essential lesson regarding sustainability for change makers: to make the world a better place we also need to tend to our inner world, learning to mend the distinct broken parts we carry within.

My healing journey was, in fact, a literacy training of sorts, helping me understand my own emotional patterns and offering insights into otherwise unconscious behaviors and choices. Seeing the value of developing greater self-awareness and self-regulation skills, I became curious: Why wasn't *emotional literacy* integrated into the work we were doing, not only in the social change sector, but in society at large?

I began loosening my attachment to my ideological roots of anti-racism and began exploring other avenues of thought, integrating research and strategies from emotional intelligence, social psychology, neuroscience, and implicit bias, as well as meditation and mindfulness. This searching would eventually result in creating a new approach I called Deep Diversity, a practical method integrating science, psychology, *and* politics to help us deal more effectively with issues of racial difference. Because of the successful impact of Deep Diversity across so many sectors and contexts, I wrote this book, which bears the same name, as a way of synthesizing the approach and lessons learned.

With its focus on compassion and neuroscience, Deep Diversity offers a different, accessible entry point into the world of racial justice. I've written the book with two audiences in mind. The primary group includes mainstream, well-intentioned people—both white and BIPOC—who recognize that racial injustice is an issue but don't fully understand it or can't "see" how the patterns of systemic discrimination play out in society.

The secondary audience is racial justice educators and activists who:

- are frustrated by the resistance and fragility they experience among people they hope to bring on board in their professional, personal, or community lives;
- have experienced, or want to avoid, emotional burnout, and want a way of doing the work that is more personally sustainable;
- have felt that traditional anti-racist/anti-oppression approaches offer useful strategies but also seem imbalanced, one-sided, or ineffective.

The opening for both of these audiences is to recognize that it is possible to effectively tackle racism, to be accountable while also minimizing shame or blame.

What I've come to understand, and what this book will explore, is that the problems regarding issues of racism specifically, and Us versus Them broadly, are not cognitive in nature—we can't just *think* our way out of them. There is no shortage of good *ideas* about how to embrace diversity, to achieve racial justice, or to make all people feel like they matter and belong. The blocks exist at a *feeling*, unconscious level. As this book will demonstrate, when we encounter those racially different than ourselves, our unconscious, emotional selves can take over.

There are both nature and nurture forces at play, a dance between biology and sociology. At the most basic level, readers will be encouraged to become aware of *patterns*, events that repeat themselves across time, institutions, and cultures. For those readers newer on the justice, equity, diversity, and inclusion journey, you are encouraged to expand your *racial pattern recognition* skills beyond hate crimes and overt acts of racism to include subtle, systemic forms of discrimination. Those with more experience, like racial justice educators and activists, are encouraged to focus on developing *psychological pattern recognition* skills, understanding the neurobiological and psychological factors that influence human behavior.

Most of the traditional approaches used to tackle racism and oppression tend to be very cognitive, or head-based, strongly influenced by a historical-sociological analysis. This is an important lens as it sheds light on the racial patterns from the past that got us to the racism and oppression we experience today. Yet it is only part of the story and does not fully explain *why* we humans do what we do. To understand the paradox of how we can be brilliant cooperators while also being brutal savages in conflict requires an understanding of our hardwired biological impulses and psychological tendencies.

After twenty-five years in the field, I'm convinced that the problems we are trying to undo are predominantly emotional in nature, so trying to *think* our way through is ineffective. Like throwing a fire extinguisher to a drowning person, it's the wrong tool for the task.

The reality is that awareness of both our inner, psychological world and our outer realities is needed to understand oppressive systems like racism or sexism. It's critical to use our heads to examine racial or gender patterns in socioeconomic data and political history, and to tap our hearts to recognize the universal human dimensions underlying them. Not doing so can result in us becoming like those we oppose, like those we wish would act or think differently regarding issues of race, ethnicity, and identity. History has shown us repeatedly that when revolutionaries are finally victorious in their fight against injustice, they most often become like the oppressors, just with new outfits.

We all have parts of our behavior, choices, and motivations that are hidden not just from other people but from ourselves. This is the domain of unconscious impulses and neural processing. In organizations, this is why so many leaders *unknowingly* discriminate despite their good intentions. The result is that the ideas and bodies of white people and men tend to get promoted and advanced. But it doesn't have to be this way.

Awareness of these tendencies—within institutions and ourselves—is key because what is hidden has more power and influence. But what we can identify and name, we can also tame. This is one of the central beliefs of Deep Diversity.

Whether you are reading this for yourself or as an organizational leader, Deep Diversity can offer valuable insights to support conversations around kitchen or boardroom tables, with family, friends, or colleagues at work. The book and related trainings have effectively supported organizations on their diversity, equity, and inclusion journeys. On interpersonal levels, it has also helped open lines of communication between self-identified Republicans and Democrats on the thorny issues of race and identity in the U.S. It *is* possible to bridge divides between people and nurture understanding, even in polarized contexts.

To do so, we must emphasize relationship-building and pattern recognition rather than dogmatic political ideology as the foundation for democracy and intergroup racial harmony. From teaching Deep Diversity extensively, I've found it nurtures an internal motivation, making people curious about themselves and their dynamic with others—a key step in reducing the Us versus Them dynamic and expanding our sense of "We." Deep Diversity is fundamentally hopeful even as it tackles the painful issues of racism, discrimination, and political division. And in this area, as in other parts of our lives, having an optimistic perspective is very important. In the words widely attributed to Holocaust survivor Elie Wiesel:

> *Hope is like peace. It is not a gift from God. It is a gift only we can give one another.*

one

The Four Pillars of Deep Diversity

What We Say, Not What We Do

A university student, Nina,[1] sits patiently in the waiting area of a nondescript office. Two other students, one Black and one white, are also waiting to be called in. After a few moments, the Black student notices his cell phone is missing and heads to the adjacent hallway to retrieve it. On his way out, he accidentally bumps the white student's leg. No words are exchanged, but once the Black student has left the waiting room, the white student mutters a nasty racial slur, one that specifically targets Black people.[2]

Nina (a fictitious student whose identity we'll discuss later) has arrived at this office to be part of a research project. She doesn't realize, though, that the study has already begun in the waiting room. What's happening is part of a Canada-U.S. study conducted by researchers from York University, University of British Columbia, and Yale University.[3] The Black and white students are actors, and the focus of the study is on Nina's response to the ugly racial insult.

Importantly, there were three research groups. One group saw this exchange happen on a video (Watchers). A second group only read about it (Readers). The third group (Experiencers) actually experienced the interaction directly with live Black and white actors.

Unsurprisingly, when asked to imagine themselves in this situation and how they might feel, the Readers and Watchers indicated that they would be outraged. When asked which student they would choose to work with in a follow-up activity, more than 80 percent of the Watchers said that they would choose to work with the Black student over the white student.

Similarly, about 75 percent of the Readers said that they, too, would choose the Black student as a partner.

None of these results should be surprising. After all, they took place in a university setting in Toronto, one of the world's most multicultural cities.

But what were the results from the Experiencers group? How did Nina and others like her feel or respond to the situation? How many of the Experiencers said or did anything in response to the racist comment?

We would expect the numbers to be a little bit lower. For most of us, responding in real time is more difficult than an imagined intervention. We might expect that maybe 50 percent of the study group would have stepped in. But perhaps that still sounds too high. A conservative guess might be that 30 percent—three out of every ten students—would have said or done something in response. Or skeptics in the crowd might suggest that only one out of ten students would step in.

The actual results? According to study coauthor Kerry Kawakami, of those who experienced the racist event firsthand, no one intervened or said anything.[4] Nor, when interviewed later, did anyone report being upset by the comment. And disturbingly, most of the students chose the white person who made the racist comment as their partner for a later assignment.

Excuse me? Yes—you read that correctly. The vast majority of the students—over 70 percent—chose the white student rather than the Black student as a partner, despite having witnessed the incident firsthand.

Here's another thing—all of the students who participated in the study were diverse in their background (except none were Black, as the study was examining anti-Black racial attitudes). On top of being well educated, young, and living in a profoundly multiracial city, the composite character of Nina could have been white, Chinese, Indian, or some other ethnocultural racial identity. A diverse bunch of university students—the odds don't get much better for a group we would expect to have empathy for their peers and potentially intervene in such a situation.

This study investigated emotions and behaviors in the context of racial difference. Although the original study took place before the powerful influence of the Black Lives Matter movement, follow-up research conducted in 2017 still found similar results.[5] The general conclusion of the researchers stands solidly: we are poor at accurately predicting how we will *feel*—and therefore react—in future situations, especially regarding racial bias and discrimination.

So why are emotions important? It turns out that how we feel directly influences how we act.[6]

Our emotions are invisible and controlling. Whether we're aware of them or not, they significantly influence our choices and behaviors. Some scientists even argue that we feel rather than think our way through the world.[7] Further, social pain (for example, from being excluded) and physical pain (from being hit, say) share overlapping neural regions in the brain. This helps shed light on why we react strongly to rejection or others' anger.[8]

To tackle contemporary discrimination and racism, we need to connect what we feel with what we think, the choices we make with how we behave. Developing emotional literacy, therefore, is the first pillar in Deep Diversity.

He Who Hesitates

Pleased with having selected a new pair of eyeglasses from a trusted shop, a consultant realizes he needs an updated prescription. The owner of the eyewear store recommends a local optometrist who does eye tests. She hands over a business card.

The consultant looks at the plain, unimpressive business card. He reads the name: Abdeiso Kiyanfar. And then he hesitates, suddenly uncertain about the recommendation. An image arises of an older, unskilled "foreign" man in a musty, disorganized office. The consultant puts the card away and heads home, with the better part of a day passing before he recognizes how unfairly he's been feeling toward the optometrist. This was a referral, after all. He pushes aside his hesitation and phones to make an appointment.

This story about a hesitation—an unstated manifestation of prejudice—is based on a true event, and it has some interesting plot twists.

First of all, the consultant in the story is a seasoned veteran in diversity and anti-racism issues. Second, his ethnicity is South Asian—he's a brown guy. The third interesting fact is that he is me.

I share this story to illustrate how vulnerable we all are—vulnerable to prejudice, racism, and bias. Also, because it holds some deep lessons about discrimination and inclusion. I am certain that if the "plain, unimpressive" business card had said Adam Wright or Ellen Goldstein, I would not have hesitated. And I would not have needed a referral, either, to take a chance on an unknown quantity with the "right" name. It's my hesitation—a brief moment of inaction—that's the issue.

Imagine if I were a hiring manager reviewing resumés and hesitated in the same way because a nonwhite name like Abdeiso Kiyanfar (not his real name) evoked a negative response? Or if I was a landlord renting an apartment and was turned off by "foreign-sounding" names? We all threaten fairness when such unconscious reluctance or preferences guide our decision-making processes when relating to others.

Project Implicit, a collaboration between Harvard University, the University of Virginia, and the University of Washington, would describe this hesitation—this first negative association with a "foreign-sounding" name—as my implicit or unconscious bias.[9] (An interesting thing, given that my name, Shakil Choudhury, would also be considered outside the norm by most North American standards.)

This is the second important insight regarding issues of diversity and inclusion. As humans, we all have biases we are not aware of that play out on a daily basis.

According to Mahzarin Banaji, one of the great minds behind Project Implicit, our implicit biases exist not on a conscious thinking level, but entirely on the unconscious, emotional plane.[10] Implicit bias may be invisible to us, but it is obvious to those who are impacted by it. Multiple studies demonstrate that there is a link between unconscious prejudice and our behaviors.[11] Consequently, understanding and uncovering implicit bias is the second part of the Deep Diversity lens.

Not till You Drink like Us

In a suburban setting, a man sits quietly watching a video. The images on the screen are of a mundane, repetitive nature: another man drinking a glass of water. The only variation is that the person shown drinking the water occasionally changes, from a white man to a Black man, to an East Asian or South Asian one.

During this unremarkable experience, something unusual starts to happen. Unbeknownst to the watcher, his brain responds differently to each image that he observes. An electroencephalograph (EEG) machine monitoring his brain activity indicates to the researchers in the University of Toronto Scarborough lab that he has greater empathy for those who share his racial background.[12]

When he watches a person of his own race, the motor-cortex area of his brain lights up as it would if he were doing the task himself. But when the person on the screen is of a different race, there is hardly a blip in the register. In fact, for some participants, observing someone of a different race having a drink of water resulted in their brains registering "as little activity as when they watched a blank screen."[13]

Thus, the third insight on diversity and inclusion. We have greater empathy—more care and concern—for those who are most like ourselves. It is widely reported in research that there are neurological underpinnings to same-race empathy as well as racial anxiety toward other-race people.[14] We are very group-ish in nature, making *identity* the third pillar of Deep Diversity. Belonging to groups not only is a key driver of human behavior, but also helps form our sense of who we are.

Our group orientation helps explain the extreme polarization between liberals and conservatives, especially in the context of the United States, and why race/ethnicity is such a strong undercurrent. The identity-based nature of politics is why, for example, so many conservatives that opposed Donald Trump's candidacy when he was running to lead the party supported him once he was chosen, in spite of the litany of scandals, lying, incompetence, and corruption that was so clearly and extensively documented. Identity was, and is, a defining feature with both nature and nurture forces at play.

But progressives also play a part in political division—even if it is a lesser role—with an overreliance on strategies that alienate, diminish, shame, blame, "call out," and push away potential allies. Highly academic jargon, self-righteousness, invisible codes of conduct, and the politics of purity often get pushed forward in the name of social justice, making it difficult to talk about issues for fear of ostracism and punishment, and further reinforcing Us/Them dynamics within and outside progressive organizations and movements. These responses, too, are tied to our being hardwired to define—and demand—in-group allegiance based on whatever social identity markers are relevant to the context, be they race/ethnicity, nationality, religion, or political affiliation.

Beyond the Third Dimension: Power

Emotions. Bias. Identity. Any element of this trio offers a formidable challenge to nurturing diversity, inclusion, and fairness in society. But it gets even more complicated. The unconscious influence of these three psychological dimensions fuses with the legacies of history, politics, colonization, and economics creating—and perpetuating—imbalanced power structures in society. The result is the entrenchment of historically high-status (more power) and low-status (less power) groups based on social identities such as race, gender, class, sexual orientation, ability, and religion.

Power and group status can sometimes be obscured in Western democratic societies like Canada, the United States, or the United

Kingdom. Here, fairness and equality are highly valued, and overt forms of discrimination, especially on the basis of race, are no longer acceptable in the mainstream society. Yet subtle, less visible forms of bias are still pervasive across institutions. They are intimately linked to the way that power in society is divided along racial lines.

For example, few white people are likely to know that their names give them an invisible advantage in the job market. But in fact, those with white-sounding names are 40 to 50 percent more likely to receive a callback for a job interview than applicants with Black or Asian-sounding names. A light skin color also increases one's chances of receiving better health care and decreases the likelihood of being a victim of police shootings. Studies that we'll look at in upcoming chapters demonstrate that this type of discrimination is so prevalent that it has to be called "systemic." And, invisible privileges as well as drawbacks come with belonging or not belonging to the racial norm in society.

This is the final insight offered by the Deep Diversity approach. Even in egalitarian, democratic societies, *power* needs to be named, challenged, and equalized to create greater fairness between racial groups.

Unlike old-fashioned symbols of bigotry such as burning crosses and white hoods, this form of systemic racial discrimination is subtle and difficult for most of us to discuss. There's a simple reason: it is hidden. It requires talking about data—the collection and analysis of the experiences of thousands of people. Such data is abstract and inherently depersonalizes the conversation for some people, while activating—at times enraging—others because it verifies their experience. And this hidden racism, when named, can further polarize the Us/Them dynamic as it identifies which racial groups have more social power and less.

The racial divide quickly gets exposed following any number of tragic high-profile deaths. In 2020, George Floyd, a Black man, was murdered on camera by a white police officer, Derek Chauvin, who choked him to death with a knee on Floyd's neck for nine minutes and twenty-nine seconds. The nation erupted with protests and outrage that spread across 150 cities, as this was another example of the prevalence of anti-Black racism.

Yet, in spite of the evidence, top government leadership in charge of the situation, Ken Cuccinelli, denied that race was a factor, instead portraying Chauvin's behavior as that of a "bully."[15]

This pattern has played out repeatedly. In 2014, a similar thing happened after the death of another young Black male, Michael Brown, who was shot by a white police officer in Ferguson, Missouri. Weeks of protests by Black people (and their allies) followed, clearly articulating the alienation and persecution they experience. The white mayor of Ferguson, in contrast, said he was unaware of any racial frustrations before that tragic shooting that made them the center of global headlines.[16]

The same sentiment emerged following the 2012 shooting death of an unarmed Black teen, Trayvon Martin, who was killed by a twenty-eight-year-old gun-toting man in Sanford, Florida. A local newspaper editor, who is white, is quoted as saying: "Until this latest incident, race didn't seem to be a huge issue in Sanford.... Even now, after it's happened, I still think Sanford is a non-racist town."

The head of a Sanford diversity organization, a man of color, responded by saying, "I understand white folks saying that... but there are Blacks in this community who have lived through [racism] a long time."[17]

FIFTY YEARS AFTER the civil rights movement, the essence of the problem we struggle with in the United States and Canada is still captured by these two statements: Race isn't an issue, says the white person. Racism is a problem that's always been here, insists the person of color.

Us. Them. On one side are millions who say they are experiencing a major difficulty that not only limits their ability to go after their dreams and get ahead but also threatens their physical lives. On the other side, millions of people say they don't see the problem and question whether it's actually "that bad." Frequently, they suggest that there may be other explanations.

Of course, no group sees any issue uniformly. People of color disagree with one another; whites disagree with whites. Yet public polls still demonstrate a significant racial divide consistently exists on the topics of discrimination and prejudice in society.[18] Although there has been an important shift in public attitude following the George Floyd protests, with 74 percent of Americans recognizing that broader racial injustices underlie such incidents—a whopping 31-point increase from 2014 following the shooting of Michael Brown[19]—how far that understanding goes still contains significant gaps. For example, a YouGov poll conducted in the months following the George Floyd protests showed that 49 percent of whites supported the protests, compared to 77 percent of Blacks. Similarly, 87 percent of Black people believed that in difficult situations, police officers would use excessive force with Black people while only 49 percent of white people believed the same thing. Overall, whites have much more trust in the institution of policing, at 56 percent, versus only 19 percent of Black people feeling the same way, a 37-point difference.[20]

In the context of political polarization, race becomes another wedge issue dividing liberals and conservatives, with the cycle of Us/Them on continual, predictable repeat. In cases like George Floyd's or Michael Brown's, as the rhetoric and pain increase on both sides, so do the extreme positions and hurtful words: *Racist! I hate white people. White justice doesn't serve us. Race-baiter! Blacks are criminals and deserve to be racially profiled...*

Besides those already mentioned, a sample list of those whose mistreatment or death have prompted racial division and heartbreak includes J.J. Harper in Winnipeg (1988), Rodney King in Los Angeles (1991), Faraz Suleman in Toronto (1996), Amadou Diallo in New York (1999), Fredy Villanueva in Montreal (2008), Eric Garner in Staten Island (2014), Sandra Bland near Houston (2015), and Josephine Pelletier in Calgary (2018).

The tragedies continued through the COVID-19 crisis in 2020, as a disproportionate number of Black and Latinx people were targeted for social distancing infractions by police.[21] In more extreme cases, Ahmaud Arbery was murdered while jogging through a white neighborhood in Georgia;[22] Breonna Taylor was shot by Louisville police in her apartment;[23] two Métis men—Jake Sansom and Morris Cardinal—were shot and killed in an altercation with a white farmer in rural Alberta;[24] and Joyce Echaquan, an Indigenous woman, filmed a Facebook Live video of herself dying in a Quebec hospital while experiencing racial abuse from staff.[25]

These are just a handful of cases from an enormous pile of evidence that racial profiling exists, within and beyond the criminal justice system. In Canada alone, over fifteen official government reports since the 1970s have shown this.[26] In the U.S., racial profiling came into public consciousness because of legal cases in Florida and Maryland in the early 1990s, with the first official legislation passing in Congress in 1999 (but failing in the Senate, therefore not becoming law). But the ugly truth is that racial profiling started with "slave patrols" in the 1700s and early police departments in the 1800s, whose job it was to control a "dangerous underclass" that included enslaved people, immigrants, and the poor.[27]

To say that historically marginalized communities, like Black and Indigenous Peoples, may feel racial profiling is a topic that's been known and studied to death would be an understatement. Yet, as a public conversation, it's a train wreck. We talk about it like we're unruly beginners.

And it's not entirely our fault.

I've come to believe that to understand what's going on, we must go deeper than skin color, both literally and figuratively. Our core struggles regarding Us/Them lie hidden in the architecture of our brains. The way that we are neurologically wired is then weaponized through the ordinary processes of being socialized into the power dynamics between racial groups. Both nature and nurture forces are at play. While we can't change our biology, we are able to alter how we are socialized, and that's our leverage.

It's important to recognize that emotionally charged issues like racial profiling and partisan politics easily trigger an Us/Them dynamic because much of what's happening is below the radar of awareness, hidden even from ourselves. Our state of "autopilot," that is, our default position—what we can't see—has great influence. And it's often unhelpful. As a result, situations, issues, and choices repeat themselves over and over. And it doesn't have to be this way.

A Compassionate Approach to Tackling Contemporary Racism

> *Once you know why change is so hard, you can drop the brute force method and take a more psychologically sophisticated approach.* JONATHAN HAIDT, social psychologist, NYU Stern School of Business[28]

For twenty-five years now, I've been an educator and consultant on issues of diversity, inclusion, and racial justice. I help organizations work through their differences, to nurture environments where all people feel like they matter and belong. Our team at Anima Leadership consults with executive teams and boards of directors when a human rights settlement requires an intervention. We train staff and students when a school has a racist incident. Our team has also developed curricula and measurement tools for public and private sector organizations, helping improve their diversity outcomes. Internationally, I've led intercultural dialogue projects for communities in conflict, specifically in Europe and South America.

For a long time, I believed that issues of racism and discrimination were simply a matter of ignorance. I thought that if we, as good citizens in egalitarian, democratic societies, had the "right" information, we would make better, more thoughtful, and fair choices. Gradually I discovered that this appeal-to-the-head-and-behavior-will-change strategy—a cognitive approach to social change—only works in a limited manner.

As the opening vignettes indicate, the problem is much more complex than being misinformed. The Yale–York–British Columbia study demonstrated that even young, educated, and ethnoculturally

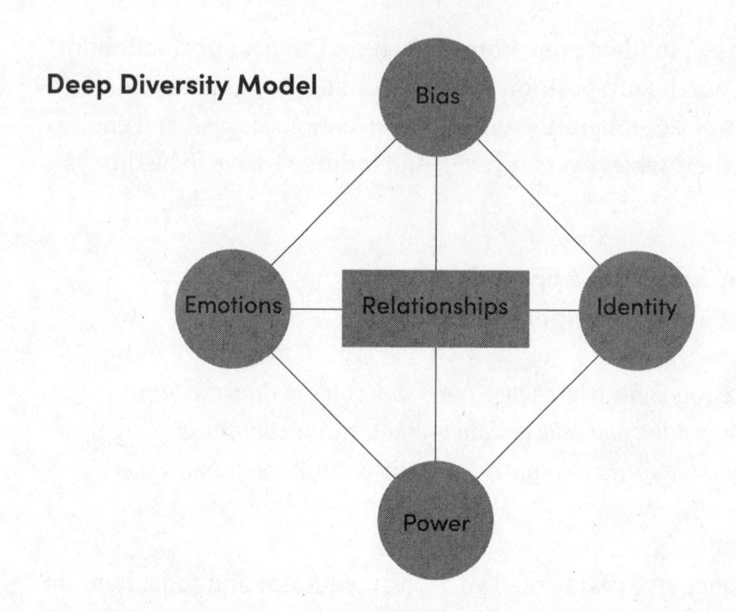

diverse students have a significant discrepancy between what they think and feel about discrimination, and what they actually do in the face of a racist event. The drinking-water study from University of Toronto suggests that we have greater empathy for those who are racially most like ourselves. And my own story about hesitating at the optometrist's name shows the presence of unconscious bias even in someone whose life's work is to educate others about issues of discrimination and difference.

An irrational preference for a white over a Black student. Greater empathy for those of our own race. A hesitation.

Notice the subtlety of each of these three examples. There are no neo-Nazis, no segregation, and no deliberate persecution. In short, there are no signs of our ugly and easily identified villain, overt racism. Yet all three examples are manifestations of its slyer but still toxic twin, subtle racism—also referred to as systemic racial discrimination. This form of racism is hard to see *unless* you have been trained to notice the patterns. Because it's hard to solve a problem we can't identify, there is a need to develop *pattern recognition* skills to identify these less obvious forms of racism, something we'll explore later in this chapter. In short, this book is an encouragement for readers to become system thinkers about race and identity issues.

As exemplified by the public conversations following the deaths of George Floyd, Breonna Taylor, or Joyce Echaquan, the inability to see each of these as reflective of system-wide racial profiling trends rather than isolated tragedies can easily fracture groups along lines of identity. And in this highly divisive post-Trumpian, post-pandemic era, this is what's holding us back collectively. This book will argue that in order to tackle Us/Them dynamics, we must first understand how central racial identity and discrimination are in the conversation. Deep Diversity offers a nonjudgmental, compassionate approach to nurturing social cohesion in both organizations and communities at large.

In short, the Deep Diversity approach helps expose hard-to-see intergroup racial dynamics. In doing so, it makes both our brain (cognition) and heart (emotions) vulnerable to constructive change. It involves learning to ask four key questions when faced with issues of ethnocultural and racial differences:

- What are the influences of emotions?
- What are the influences of bias?
- What are the influences of identity?
- What are the influences of power?

Using Brain Science to Understand Us/Them

At the heart of the Us versus Them racial dynamic is our tendency to see a person as a symbol of a group, rather than as an individual.[29] When we do this, our empathy is reduced and we may dehumanize the person in some small or big way. Research described as "robust" demonstrates our tendency to see those who are racially different in simplistic, primitive stereotypes—more like animals or objects than people.[30] This tendency, called infrahumanization or objectification, shows itself whenever we make generalizations about a subset of people, especially so-called minority groups. (For example, making derogatory associations of Black people with apes and violence, or seeing East Asians as hardworking, expressionless, and almost robotically efficient.)

Fundamental to this discussion is understanding that our unconscious mind—which is automatic, reactive, emotional, and intuitive—easily dominates the conscious mind, the realm of logic, language, reason, and abstraction.[31] In the words of a respected researcher, Joseph Ledoux of New York University, "consciousness may get all the focus ... but consciousness is a small part of what the brain does, and it's a slave to everything that works beneath it."[32]

As this book will show, our unconscious biases and automatic brain processes frequently favor those most "like us," creating hard-to-see racial discrimination that becomes systemic against "them." These factors coupled with power dynamics and group status function as the root system of discrimination, preventing many hardworking individuals and communities from getting jobs or accessing appropriate health care services, or causing them to be overpoliced. This same combination of biological impulses and social conditioning contributes to the divisiveness in politics.

Turning inward to the level of gut feelings and emotions will offer a greater appreciation of the problem, as well as possible solutions. As we appreciate the depth of what we collectively face and are able to generate some compassion for why we get stuck, we may be better positioned to progress toward racial equality and political cooperation in new, more helpful directions.

In brain science terms, we have to disrupt and alter the neural pathways that result in biases that do not serve us collectively.[33] In plain language, we have to break some bad habits regarding issues of racial difference.

Breaking Bad Habits; Nurturing Good Ones

To say that we, as humans, are creatures of habit is a profound understatement. By habit I'm referring to something much bigger than a daily ritual of grabbing a coffee before work or going to the gym at set times. Most of our waking life is habit-based, from our posture, walking gait, manner of speaking, accent, and favorite expressions to how we think, chew, eat, feel, laugh, move, play, or sleep.

A significant part of our lives, therefore, is experienced in a state of autopilot. We don't think about what we do, we just do.[34] This is both the brilliance and the drawback of our neural architecture. For example, learned behavior that becomes unconscious (such as driving a car) allows us to function efficiently. It frees up our mind to pay attention to new, less familiar things in our environment (a driver in the wrong lane) and prioritize our actions (swerving out of the way).

If we had to pay conscious attention to all the stimuli in our environment at any given moment (all the minor and major tasks required in driving a car), even basic events would be overwhelming. We couldn't focus on what was important, which could put us in harm's way. The flip side is that such autopilot results in filters and bias, and a state of comfort with how things are. This makes change difficult for us.

Habits are also larger than the individual. Habits shared by a group of people—whether in a corporation, profession, or nation—become social norms (also called cultural norms). Our environment marinates us in these customs, behaviors, and thinking patterns. They range from what side of the road we drive on to how we greet people (with handshakes or kisses), how much eye contact we make (as a sign of respect or disrespect), and what kinds of clothing or public displays of affection are acceptable. We are social beings whose individual identities are influenced by a wide variety of forces, including ethnic and gender norms, family, friends, coworkers, communities, social class, and religion, to name just a few.

A large body of research demonstrates that many social norms are invisible (for example, greeting a stranger with a handshake) until they are breached (being greeted with kisses on both cheeks by someone you don't know).[35] And the experience of norms being broken can often be emotional. We might have a sense that something is not right or feel confusion, embarrassment, or anger. As a result, we may begin to avoid specific people, places, or situations.

Whether such habits are learned or genetically influenced, they are usually unconscious. Our culture, then, provides the "software" that both helps and hinders racial interactions. But Deep Diversity also explores the "hardware" side of things. It turns out that we are biologically predisposed to bias and discrimination. We are preprogrammed from birth to define who are our "people," who is part of my in-group and who is not.

This aspect of the research demonstrates that bias, fear, and prejudice are intertwined in the mind with normal processes of how we perceive, categorize, remember, and learn.[36] It suggests that the mechanisms for Us versus Them are, at least partially, built into the human design. From my experience, accepting and admitting this is a good collective starting point. It gives us permission to have conversations about bias, prejudice, and discrimination that otherwise are almost taboo. After all, how can we talk productively about prejudice and discrimination if these concepts are associated only with neo-Nazis and overt racists but not with the rest of us?

Enhancing diversity, inclusion, and equity requires both learning and unlearning about others and ourselves. It requires developing more psychologically sophisticated strategies to advance diversity and inclusion. In short, it requires us to break generations of bad racial habits and nurture good ones. Unless we understand the roles played by both hardware and software components in our mental processes, fairness and justice will always be threatened in our personal, professional, and community lives.

The Deep Diversity Model

The Deep Diversity approach, then, argues that we must understand the brain and mind—its conscious and unconscious dimensions. The psychobiology underlying human behavior motivates racial interactions, as do socialization processes. On one level, when faced with issues of racial and ethnic differences, Deep Diversity asks: *What are the influences of emotions, of bias, of identity, and of power*?

On another level—if you can imagine the mind as a computer—it's about disrupting and debugging the Us/Them default setting that the human program is predisposed to. Attempting to answer the four questions posed by the Deep Diversity approach makes possible a more nuanced understanding of what's happening for, and to, us.

Each of the four key questions can be a gateway into a series of related questions. For example, the question of *emotions* in a racial situation, which Chapter 2 focuses on, may generate the following:

- What do I feel about this situation, group, or issue?
- What do others who are involved in this situation, group, or issue feel?
- What is the emotional history or tone underlying this situation, group, or issue? And how might that history still influence perspectives and outcomes today?
- How might we address these feelings and the underlying needs?

Similar questions need to be generated on the topics of bias, identity, and power. Chapter 3 demonstrates that all humans have implicit biases that result in personal and societal deficiencies that go unnamed regarding issues of racial difference. Such invisible prejudices lead to behaviors that can undermine fairness and impair judgment, with outcomes that can be mild (for example, social slights) to severe (violence and death).

Chapter 4 explores the concept of identity. Belonging to groups is a significant driver of human behavior, as important as the need for food, water, and shelter. Our sense of who we are as individuals is developed through our interactions in groups both consciously and unconsciously.

Understanding our group-ish nature will help us explore the divisive nature of politics today from another angle. A look at social identity will help us explore why, for example, Republicans and Democrats see each other as enemies today rather than adversaries, and how that distinction is a threat to democracy itself.

Chapter 5 focuses on power, exploring the interplay of social power between dominant and nondominant groups. This is a critical lens in tackling systemic forms of discrimination such as racism. Power is a complicated subject, though, that requires more than a single chapter.

Chapter 6, therefore, explores personal power, our leverage in helping achieve racial justice. This section argues for developing resilience by beefing up our psychological and spiritual qualities important to advancing equity and social cohesion while preventing burnout. The conclusion of the book, in Chapter 7, brings all the pieces of the Deep Diversity approach together.

NEUROPLASTICITY: CHANGING HABITS OF THE MIND

Deep Diversity focuses on habit breaking and forming. It is entirely possible to change our habits: there are proven ways to change how we think and perceive, thereby developing new neural patterns. There is considerable evidence that human brains exhibit neuroplasticity, meaning that where we put our attention and what we focus on helps change the structures of the brain.[37] Just as muscle mass can increase through physical workouts, we can grow parts of our brain, allowing us to form new habits and shift our perspectives in the ongoing task of increasing racial harmony.

Many of the strategies outlined in this book build on the principle that thinking influences actions, similar to what is taught in cognitive-behavioral therapy (CBT). CBT is an evidence-based form of treatment that has been used since the 1960s to help people change unproductive ways of thinking and behaving.[38] CBT has proven successful in helping people address a wide variety of issues, including depression, anxiety, eating disorders, anger management, post-traumatic stress disorder, and low self-esteem.[39] Why not consider similar principles to tackle prejudice and stereotypes? After all, biases and preconceptions about others are, in many ways, just distortions in our thinking. Change the thinking, change the behavior.

INNER SKILLS AND COMPASSION SUPPORT DEEP DIVERSITY

In the words of neuropsychologist Rick Hanson, "you can do small things inside your mind that will lead to big changes in your brain and your experience of living."[40] Deep Diversity emphasizes that we can change how we think and feel about others and ourselves, especially regarding racial issues, so as not to be unconsciously swept away by ancient brain processes and structures. It requires some work and effort, but it is entirely possible.

These "small things" that Dr. Hanson refers to I'll call inner skills, as they have to do with what's going on inside our individual heads and bodies. These are microabilities that help us stay balanced and centered when things get emotionally heated. We can call on them when controversy clouds our thinking and threatens to divide people along lines of race, ethnicity, and identity.

Each chapter will introduce one or two inner skills to enhance positive relationships with ourselves and others. The skills—self-awareness, meditation, self-regulation, empathy, self-education, discernment, relationship management, conflict competence, and meaning-making—serve as the action component of the Deep Diversity model. These skills can be developed and enhanced. And they can help us nurture greater ease, resiliency, and flexibility in dealing with others.

To support systemic change efforts, it's also crucial to develop enhanced awareness of systemic patterns. I'll refer to this as *racial pattern recognition* (RPR), the ability to identify repeating trends and cycles in society—and within ourselves—in order to expand our knowledge base to include subtle and system-wide forms of racism and discrimination, not just the overt. Understanding patterns helps make meaning of the world around us and is at the heart of the human experience, whether it's children decoding the patterns underlying reading and mathematics or a skilled doctor quickly detecting a life-threatening illness from a patient's symptoms. Expanding our RPR repertoire to include, for example, that resumés with nonwhite names are often more critically assessed may help us counter the pervasive invisible pro-white bias and enhance fairness in organizational decision-making.

It's also important to develop *psychological pattern recognition* (PPR) skills, a term I'll use to refer to the neurobiological tendencies influencing the formation of our social identities and how we navigate the world. Returning back to the resumés example, PPR helps us understand that while we work to eliminate pro-white bias and racial oppression, it's important to avoid concluding there is something inherently "bad" or "wrong" with white people. As we will explore through this book, there is a universal tendency—both unconscious and conscious—for the dominant social group to design the rules and norms within a society to most benefit their own group members, resulting in mistreatment and unfair power dynamics with minoritized groups. Understanding such global patterns may support the development of more effective strategies to disrupt discriminatory behavior, and increase the participation of nondominant people in decision-making while also motivating dominant group members to support change efforts without an overreliance on shame-and-blame tactics.

Therefore, at the end of each chapter, you will be encouraged to reflect on, and expand, your RPR and PPR repertoire with a dual focus on systemic patterns and psychobiological tendencies.

Compassion, the "mental state of wishing that others may be free from suffering," is the underlying principle for all of this work.[41] It is needed because diversity, by its very nature, is about our differences. It is about the places where we are not, and may never be, the same. We will—we must—make mistakes, both big and small. That's part of the process of learning about each other. But making mistakes feels bad, so most of us have learned to avoid, deny, or minimize them.

When walking through the field of racial justice and discrimination—which is pebbled with mistake-making opportunities—our "secret power" is compassion. Turned inward, self-compassion can help soothe the voice of the inner critic that judges harshly and tends to dwell on errors and missteps, preventing us from moving forward. It can help us all accept that humans—regardless of background, color, or identity—are imperfect creatures. We are all on a learning curve regarding unconscious prejudice and intergroup dynamics. Accepting this as part of the process may help us move past the internal and external judgments that hinder learning.

As we begin to grasp the impersonal nature of prejudice and discrimination—that it's a leftover of primitive elements of our brain—we can generate more compassion for others and ourselves. Some of our struggles are a result of neural hardwiring, shaping a good amount of our group-ish nature. However, a significant amount—a result of socialization, norms, and socioeconomic structures—is changeable. The former is not our fault, but the latter is our responsibility.

INTENTIONAL FOCUS ON RACE

Although Deep Diversity has implications for a variety of equity issues—including gender, class, sexual orientation, and disability—this book focuses on race and ethnicity. This focus is intentionally narrow. It will allow us to drill down into one specific area and apply the method of Deep Diversity comprehensively. As we understand the patterns of difference more clearly and grasp the hair-trigger tendency to fracture into Us and Them, the possibility emerges of creating a map for other equity issues.

This book is steeped in my identity as a middle-class, heterosexual, cisgender, able-bodied, university-educated Canadian male of South Asian ethnicity. I was born in Pakistan, the child of two parents who survived *partition*—the bloody process of overthrowing the British that resulted in the independent states of India and Pakistan. I have family today on both sides of those national and religious divides, representing both Islam as well as Hinduism. In the early 1970s, civil war between West Pakistan and East Pakistan—now Pakistan and Bangladesh, respectively—again pushed my family into exile, this time arriving in Canada. It was about a half dozen years of the immigrant shuffle before we landed on our feet in the 1980s. I claim my identity and personal journey as the source of strength and insight for this book, and openly name it as the basis of bias and weakness, both conscious and unconscious. I don't believe we can ever be truly objective; our worldview is always tainted through our personal filters. But we can be transparent about our starting points.

I'm also not a neuroscientist or psychologist. I have, therefore, provided citations for the research to back up my claims, and stayed within ideas that are generally accepted in these fields. I've also invited a wide variety of professionals, including those with science and psychology expertise, to review my writing. The final decisions, however, are mine alone, and I take full responsibility for what's written here. Just as a professional coach in the business world knows they are not a therapist and can only work to a certain level of depth with clients, I have learned to be respectful of my professional boundaries. But I've also had to step beyond my field of expertise, developing an interdisciplinary approach that I think is important to share.

That's my entry point.

Regardless of your starting point in this conversation about diversity and fairness, the journey ahead is a shared one. It's about all of us as humans—there is no "them." The collective step, however, begins with the individual. It begins with each of us venturing out of our comfort zone and taking a risk, trying something new.

In the words of world-famous musician Yo-Yo Ma: "Things can fall apart, or threaten to, for many reasons, and then there's got to be a leap of faith. Ultimately, when you're at the edge, you have to go forward or backward; if you go forward, you have to jump together."[42]

Emotions: Understanding Ourselves and Others

Crisp Winters, Burning Women

The photos on the website are inviting. They show the kind of small-town pastoral splendor that's like catnip for city dwellers like me: flowing rivers, lush forests, kids and horses, and even a beautiful white church with tall spires. The site goes on to say that "the surrounding countryside is reminiscent of times past, lazy country summer days, crystal clear streams and lakes with cold crisp winters in an unspoiled environment."[1]

Wow—I'm in. Sign me up.

Clicking through the site, I find something called the town charter, which has been approved by the mayor and six council members. The declaration includes a section entitled "Our Women."

Hmmm, interesting. I scroll down and read:

> We consider that men and women are of the same value. Having said this, we consider that a woman can; drive a car, vote, sign checks, dance, decide for herself, speak her peace [*sic*], dress as she sees fit respecting of course the democratic decency, walk alone in public places, study, have a job, have her own belongings and anything else that a man can do. These are our standards and our way of life.

This is starting to sound somewhat unusual. What is it—perhaps the website of a commune or a retreat center with pseudo-feminist leanings? The last line, though, really throws a wrench in the works:

> However, we consider that killing women in public, beatings, or burning them alive are not part of our standards of life.

Wow. What part of the world could promise both cold crisp winters and a strong rebuke against burning women, advertised in the community's code of conduct? Confused yet?

Welcome to Hérouxville, Canada. Population: 1,300 people.

In 2007, this community in the French-speaking province of Quebec made national and international headlines when it passed the now infamous Hérouxville Town Charter. (The original charter even forbade the "stoning of women.") The context? No local or regional cases of such gender-based violence prompted the town's charter. It seemed a not-so-subtle message targeting "immigrant" groups, specifically Muslims. But immigrants (and people of color, generally) are almost nonexistent in this very white region of the country. So where was this coming from?

There were no actual experiences or problems in the region, let alone the village, which might have motivated the town charter. We are left to speculate why the town council would take such drastic steps. I'll leave that question hanging for a moment, as the actions of this small town were a harbinger of things to come on a larger scale.

In the following years, this prompted a fiery debate across Canada on immigration and racism. Subsequent governments seemed to draw pages from the Hérouxville playbook both inside and outside the province. This included Stephen Harper and his Conservative party who campaigned in 2015 on creating a police hotline to report "barbaric cultural practices," taking a hard-line stance in Harper's bid to be reelected prime minister of Canada (he lost to Justin Trudeau).[2] In Quebec, by 2019, this debate culminated in the passing of the *Laicity Act*, a bill that promoted an extreme form of secularism in the name of "neutrality." The centerpiece was banning the wearing of all overt religious symbols—including Muslim hijabs, Sikh turbans, Jewish kippahs, and large Christian crosses—by public employees including doctors, teachers, government officials, and day care workers.[3] Legislation was even passed that banned face coverings when even *receiving* public services, but this law got hung up in the courts as it so obviously targeted, and discriminated against, Muslim women.[4]

Harassment of minorities and dramatic increases of racism resulted throughout these years, including property damage and ugly confrontations with racial slurs and physical assault, and peaked with a horrifying mass shooting in a Quebec City mosque in 2017, in which six people were murdered and nineteen others seriously injured, all Muslim.[5]

On a macro level, the trends in Quebec were similar to those of Donald Trump, who took anti-immigrant sentiments to a whole new level of toxicity during his four years as U.S. president between 2016 and 2020. He described Mexicans as criminals, drug dealers, and rapists and also suggested that all Muslims in the U.S. be registered and tracked—a concept chillingly similar to how Jews were "marked" with gold stars during the Hitler era in Germany. One of his very first executive orders was to ban entry of immigrants and refugees specifically from predominantly Muslim countries, including those fleeing from the brutal Syrian civil war.[6]

Similarly, hate crimes skyrocketed under the Trump administration with impacts felt by many including Muslim, Jewish, East Asian, and LGBTQ2S+ communities. It's not a surprise then that Trump's consistent attacks on Mexicans correlated with a 41 percent increase in hate crimes against Latinxs during his tenure, including a racially motivated mass shooting in El Paso, Texas, that killed twenty-three and injured twenty-three others in 2019.[7] In fact, European researchers used data to demonstrate that Trump triggered a "global racist contagion" as hate crimes spiked upward in many nations including Austria, Belgium, Canada, Czech Republic, France, Germany, Israel, Poland, and Russia.[8] The United Kingdom was part of this upward trend, with a record-breaking 10 percent jump in hate crimes reported in 2019 alone—with 76 percent of these being race-based offenses.[9]

Although there are different circumstances and histories between the context of Quebec and the context of the U.S. under Trump, mainstream media analyses also identified some overlapping themes including the amplification of anti-immigrant prejudice, Islamophobia, and the urban-rural divide. Some political observers suggested that the divisive tactics used by Quebec's center-right government—Coalition Avenir Québec—and the Republican Party under Trump were gambles on the part of desperate political parties with aging and shrinking support bases. Mobilizing conservative, mostly rural voters was strategic, as these people had the strongest ties to the traditional, white, Euro-cultural heritage amplified by each of these political parties.

But what I find most relevant to the Deep Diversity discussion is that in this conversation, *fear* was a critical undercurrent.

But why is fear so easily triggered? When it comes to dealing with those whom we perceive to be cultural outsiders, why is it so easy to evoke feelings of anxiety, suspicion, or even panic? The emotion of fear—often present in these situations but usually invisible—opens a way to examine the broader, unseen role emotions play in our encounters with those who are, or are perceived to be, different than us.

Investigating the fear response allows us to expose a key underlying factor that powers Us versus Them, and thereby weaken it. This dynamic can manipulate us into being reactive rather than thoughtful, resulting in choices that can hurt our relationships and communities. We'll also discuss how the inner skill of self-awareness can help us identify what's happening internally, so that when it comes to issues of racial difference, we are better able to respond rather than react.

I'll focus on Hérouxville and its town charter to make this point, as this micro example more easily demonstrates the larger patterns seen in the U.S. and globally. Similar emotional dynamics play out whenever Us versus Them emerges, whether within institutions or in broader society.

Developing Emotional Literacy

> *Emotions do more than color our sensory world; they are at the root of everything we do, the unquenchable origin of every act more complicated than a reflex.* THOMAS LEWIS, FARI AMINI, and RICHARD LANNON, *A General Theory of Love*[10]

In the 1990s, psychologist Daniel Goleman helped popularize the principles of emotional intelligence. Since then, a considerable library of materials has developed on the purpose and power of emotions and their controlling yet invisible role in our lives. Developing our emotional quotient (EQ) has become widely recognized as critical to personal and organizational success. EQ is regarded by many to be as important as IQ (intelligence quotient), the traditional measure of intelligence.[11]

In daily life, emotional intelligence can be defined simply as how well we, as individuals, manage ourselves in relationships with others. This sounds deceptively simple. Most of us believe that we handle ourselves pretty well and would likely say that we are good at managing our relationships. Where a relationship is not easy, even while accepting some responsibility, we're likely to point to shortcomings in the other person. They are angry, self-centered, and insecure. It's rare that we notice how our own actions, tone, or behavior may have contributed to or instigated the problem.

Even when we pay lip service to the idea that we're not perfect, we put our energy into finding fault in the other. Through this exercise of faultfinding, it's difficult to see other people clearly. Our unconscious motivations, bias, fear, and history—our emotional baggage—gets in the way.

This is especially true regarding issues of diversity and intergroup differences. Emotions play a crucial role in the Us/Them dynamic. Feelings are at the roots of our actions, whether we are aware of them or not. A significant portion of our decision-making lies below the surface of our awareness. That's why developing emotional literacy is critical.

From a mountain of good literature on this topic, here are three ideas that are helpful for understanding the unconscious and automatic nature of emotions as they are relevant to issues of racial difference:

- **Tilting toward/away:** We are inclined to tilt toward or away from things in our environment; this is also called the *approach-withdrawal system*. Whether we are aware of it or not, we tend to tilt toward those most like ourselves and away from those we perceive to be different.
- **Emotional contagion:** The contagious nature of emotions and the open-loop structure of our nervous systems mean we are designed to regulate each other. When we feel included, we tend to soar. When excluded, we tend to underperform, second-guess ourselves, and, in extreme cases, get sick.
- **Emotional triggers:** The midbrain region, called the limbic system, modulates our emotions; the amygdala specifically alerts us to dangers in our environment. Strong emotional triggers can activate the *fight-flight-freeze* response, reducing our ability to think clearly, especially when dealing with those who are racially different than us.

Tilting Toward/Away: Our Survival Instincts

The primitive nature of our brain is well established in research: it was designed to survive physical threats and emergencies more than anything else.[12] Although that function may have suited our ancient cave-dwelling relatives who lived in small, violent groups, it can be problematic for interconnected, globalized societies in which billions of people are attempting to live together.

Like other animals, we have a very simple survival orientation: tilting either toward or away from things. Generally, we are attracted to tasty foods, pleasant smells, friendly people, warm blankets when we feel cool, and cold drinks when we're hot. At the same time, we will jump away if we think we see a snake, express disgust at rotting foods and animals, pull our hands back from a hot fire, and generally avoid erratic or dangerous people.

According to social psychologist Jonathan Haidt, both tilts are part of an elaborate mechanism to help us keep our physical and emotional states in balance.[13] The approach-withdrawal system develops from genetics as well as social and environmental factors. It also applies to our relationship to human groups based on identity: we gravitate to those who are most like ourselves, and are shy or fearful of those who are different.

There is an exception to this general rule, which we will explore further in the next chapter: that members of lower-status groups may often prefer those from higher-status groups, rather than those of their own group, because of the dynamics of power and socialization.

NEGATIVE TILT STRONGER THAN POSITIVE

It's important to note, however, that the two tilts are not created equal: the tendency to withdraw is more powerful than the tendency to approach. We have what's known as a *negativity bias*, which primes us for avoidance and remembering the bad even when it's outnumbered by the good.[14]

From my experience working with people—whether with junior high school students or with senior management teams—I find they will spend much more time discussing, for example, what went wrong at the end of a project rather than what went right. It's rarely proportional. The negatives are almost always given far more airtime than is deserved. The positives are skimmed over relatively quickly.

Neuropsychologist Rick Hanson describes this phenomenon as the brain's tendency to be "like Velcro for negative experiences and Teflon for positive ones."[15] Of course, people don't need to have advanced psychology degrees to figure this out. It's why political smear campaigns are so effective and why news reports seek to attract an audience by focusing on what's going wrong in the world. Well established in research, this negativity bias is much less overt than a conscious thought. It exists instead at the subtle feeling level. It's also closely related to fear, which is believed to be our oldest emotion.[16] As a result, we recognize expressions of fear on faces more quickly than happy or neutral ones. One brain structure related to emotions, the amygdala, is quickly activated by fearful faces even if they're not registered consciously.[17] The impact has also been shown in relationship to diversity. Greater negativity arises when we're dealing with those we perceive to be different than ourselves, especially racially.[18]

We can track this tendency back to early humans. It would have been an evolutionary advantage to tune in to danger, fear, aggression, and general negativity expressed by unfamiliar people who may have been a threat to survival.

Much research, which this chapter and the next will expand upon, demonstrates that we generally tilt toward those most like ourselves while tilting away from those who are different. This impacts our choices of where we live, work, and play and whom we choose to be part of our social networks.

For example, just northwest of Toronto in the suburb of Brampton, where my family lived for many years, a massive influx of South Asians has taken place since the 1990s. This group now makes up almost 40 percent of the population. (In fact, people of color are a majority there, making up over 60 percent of the population.)[19] Similar to Birmingham, England, many people with roots in India and Pakistan have made this place their home, drawn to a city that has a lot of people like them (a tilt *toward*). Orange County, California, has about 190,000 people of Vietnamese ethnicity, centered on a community called Little Saigon. Such ethnic enclaves—Chinese, Italian, Greek, Polish, Jewish—are extremely common and have always existed, in some variation, in most large cities.

On the other hand, *white flight* describes the phenomenon in which white people have left Brampton—and other North American city cores—in significant numbers. London, England, has been a white-minority city since 2011, with an estimated 600,000 white people having left over the previous decade.[20] Many felt uncomfortable with their place in the increasing ethnocultural diversity (a tilt *away*), preferring more homogeneous white communities and small towns outside the city (again tilting *toward*).[21]

In the context of Hérouxville, negativity bias and the tilt-away phenomenon might help us understand how the fear of difference became so easily activated, even when there was little local experience with immigrants or people of diverse backgrounds. The town charter seems to have emerged from a defensive posture, a wariness of imagined immigrants.

Emotional Contagion: Being Controlled by the Moods of Others

> *The open-loop design of the limbic system means that other people can change our very physiology—and so our emotions.*
>
> DANIEL GOLEMAN, RICHARD E. BOYATZIS, and ANNIE MCKEE, *Primal Leadership*[22]

Emotions surround all human dynamics, influencing our interactions on conscious and unconscious levels. Many experiments demonstrate that feelings are contagious. They can be transferred between people, like catching a cold.[23] Our heart rate, blood pressure, and mood, for example, are easily synchronized with others in our vicinity. Those who are emotionally dominant can transfer their mood to others without effort, prior history, or words spoken.

These effects, referred to generally as *emotional contagion*, occur with family and friends, in boardrooms, on the shop floor, or when dealing with clients and customers.[24] Emotions spread quickly and easily, influencing interactions in our private, public, and professional lives.

The physical form of our bodies can convince us that we are a series of self-contained units—*closed loops*—that are separate entities from other people. Although there is some truth to this (otherwise we would be leaking blood and fluids everywhere we went), it's also partly an illusion. Neurologically speaking, we are considered *open-loop systems*.[25]

Our nervous systems are designed to *co-regulate*, to tune in to and intermingle with each other's physiology, making neural connections. Specifically, our emotions play a significant role in our biochemical regulation. We are designed to regulate, and be regulated by, others.

By design, we are also exquisitely sensitive to social pain such as exclusion and ostracism. In the words of neuroscientist Matthew D. Lieberman, "when human beings experience threats or damage to their social bonds, the brain responds in much the same way it responds to physical pain."[26]

Lieberman's team was the first to demonstrate that social and physical pain areas overlap in the same region of the brain.[27] Kipling D. Williams from Purdue University has shown that even brief experiences of exclusion during the playing of something as insignificant as an online pass-the-ball computer game can result in strong emotional reactions. Participants demonstrated "unusually low levels of belonging to groups or society, diminished self-esteem, and lack of meaning [in], and control over, their lives."[28] Why this is important is that humans come equipped with an instinct to survive—the foundation underlying the primal drive for food, water, and shelter students are taught in basic biology class. In many ways, the experience of social exclusion promotes a visceral survival-threat response within all of us, because deep within our neural circuits is the evolutionary knowledge that to be excluded from the tribe meant a low chance of survival and increased likelihood of death.

The upside to this neural connection is that joy, positivity, calmness, and rationality can also be transferred between people. Emotional intelligence research has clearly shown that when individuals or groups feel positive and upbeat, everything tends to go better, including creativity, problem-solving, productivity, understanding complexity, and predisposition to being helpful.[29]

The downside of our ability to co-regulate is that other emotions like chronic anger, anxiety, or a sense of futility can also be transferred, damaging relationships and hijacking our work or personal environments. Generally, when we are upset, stress hormones are secreted that take many hours to be reabsorbed by the body and fade. They impact our ability to rest, sleep, and recover.[30] In a toxic workplace, for example, where conflict, distrust, or dysfunctional relations are the norm, not only is productivity reduced but the health impacts on employees can also be significant, resulting in sick leaves and absenteeism. In Canada, it is estimated that stress-related absences cost employers $3.5 billion annually, while in the United States, that figure is ballparked at $300 billion.[31] Similarly, in the U.K., stress costs businesses about $34 billion yearly.[32]

LEADERS SET THE EMOTIONAL TONE

According to emotional intelligence research, leaders play a role in how people feel. In a group, they serve as emotional guides.[33] Leaders' words and reactions carry more weight than those of other group members; they are watched more carefully and are given more eye contact. We take many of our cues from those in charge.

People in positions of authority generally set the tone for appropriate group behavior, especially in times of uncertainty, rapid change, or conflict. For example, leaders who are able to stay calm during a crisis can settle group members. A manager who demonstrates mild anxiety or hesitancy may communicate to the team that something still needs careful thought or attention. Leaders can inspire us, evoke our empathy, or fuel our patriotism and anger through a call to arms against an enemy. The quality of leadership, therefore, plays an influential role in our lives with both negative and positive impacts, including in the context of cultural differences.

Coming back to Hérouxville, there's some evidence that one of the town councillors, a strong anti-immigration activist named André Drouin, played a leadership role in drumming up support for the charter—and, likely, fear. Drouin, the key spokesperson for Hérouxville regarding this issue, was known to speak bluntly, referring to multiculturalism as "idiocy" and demanding a moratorium on immigration to Canada.[34] He was responsible for drafting the controversial legislation, and asserted that he brought this issue forward intentionally to make Hérouxville a case study in a broader international anti-immigration movement.[35]

Drouin's anti-immigrant bias was accompanied by great passion and purpose. Given that he had a significant leadership role—he was an elected town official, after all—it wouldn't be outrageous to suggest that he served as the emotional guide for Hérouxville's fearful and defensive *tilt-away* posture that resulted in the unusual code of conduct.

What leaders say and do *matters*, whether in a small town or on the global stage.

In the context of the U.S., the words of Donald Trump—his well-documented exaggerations, lies, and attacks on opponents—divided and damaged public life like no previous president.[36] From the minor (his falsehoods about the large size of the crowd at his inauguration) to the dangerous (denial of the seriousness of coronavirus and suggestions to "treat" it with bleach injections) to the catastrophic, such as his propaganda campaign about how the 2020 U.S. election was stolen when he lost to Joe Biden. Through all of this, most of his followers believed him, and the consequences are still beyond dire. Americans have become so distrustful of government, opposing political parties, and each other that U.S. democracy itself is in peril.[37]

Emotional Triggers: The Role of the Amygdala and Limbic System

To understand emotions and their origins, we need to back up a bit and revisit the brain. It's believed that our brain evolved in stages, resulting in three distinct sections known as the reptilian, limbic, and neocortex. From an evolutionary perspective, the reptilian brain is the oldest and most primitive part of us. It regulates our automatic functions such as breathing, heart rate, startle function, swallowing, and a host of other tasks that are essential to basic survival.[38]

The next region in line to develop was the limbic brain, a feature we share developmentally with other mammals. This part of our brain is responsible for the "share and care" parts of our personality. It is critical for nurturing and defending our young, as well as vocal communication, play, community, empathy, and socialization.

The youngest brain region to develop was our neocortex.[39] This is the metaphorical home of our conscious mind. Thinking, attention, abstract reasoning, fine motor skills, and language are rooted here. The prefrontal cortex, the section encased by our foreheads and behind our eyes, is particularly important. It is believed to be the brain area that determines our capacity for emotional intelligence.[40] The prefrontal cortex is responsible for a variety of executive functions including setting goals, planning, directing action, and guiding, as well as inhibiting emotions.

The home base for emotions is in the limbic region. This part of the midbrain houses many structures including the amygdala, which constantly scans for threats and is the trigger point for the body's fight-flight-freeze mechanism.[41] Rick Hanson describes the amygdala hub succinctly:

> Moment to moment, the amygdala spotlights what's relevant and important to you: what's pleasant and unpleasant, what's an opportunity and what's a threat. It also shapes and shades your perceptions, appraisals of situations, attributions of intentions of others, and judgments. It exerts these influences largely outside of your awareness, which increases their power since they operate out of sight.[42]

A perceived threat by the amygdala can set off the body's fight-flight-freeze mechanism, a survival response to fend off an incident that may be life-threatening. The automatic response easily overpowers the thinking part of our brain. In this state, we become very reactive. This can work in our favor and help us, for example, jump out of harm's way (from a car or snake, say). But it also has drawbacks. For example, the amygdala can misfire when we interact with those who are different than us. It has been shown to be activated when we relate to those of a different race, suggesting that a potential cascade of negative unconscious feelings and bias are also at play in racial interactions.[43] Recall from Chapter 1 that there already is a tendency to demonstrate less empathy for those different than us, and especially in the case of racial minorities.

Integrate into this mix our built-in negativity bias, and the outgroup cocktail keeps the brain scanning for threats, amplifying other unpleasant feelings such as anxiety, anger, frustration, shame, or guilt.[44] Because we don't seem to have an equally strong automated response mechanism for positive inputs, we have to work harder to keep track of the good things.

The result is a tendency to misjudge members of other racial groups, treating them unfairly. My experience of consulting inside organizations for many years highlights very common patterns: that members of minoritized groups—Black, Indigenous, people of color, people with disabilities, or those who identify as LGBTQ2S+—are less likely to be hired, advance, or be promoted as compared to their more normative counterparts. These groups often face greater criticism, harassment, and discrimination while receiving less acknowledgment of their successes as well as minimal professional support via coaching, career planning, or mentoring. Unless we develop pattern recognition skills to identify and disrupt these manifestations of subtle racism, thinking will often take a back seat to feeling when dealing with those we perceive as "different."

In the social contexts we've discussed, the aggressive fearmongering of leaders like Drouin or Trump fed the overactive amygdala responses, and likely played a role with the average Hérouxville resident or U.S. Republican, respectively. And why wouldn't they? There is much in the social background that primes people to feel on guard, wary, and frightened.

Much of the population across industrialized nations like Canada, the United States, and the United Kingdom have felt financially insecure due to the impacts of forty years of neoliberal economic policies. Since the 1980s, wages have been stagnant for average people while expenses have only gone up. Part-time and precarious jobs have replaced full-time employment with benefits and pensions, something that is especially true in rural areas and specific regions, whether the northeast of England or the U.S. Midwest—previously the manufacturing and industrial heartlands of each nation. Research shows that the gross domestic products (GDPs) of Canada, the U.S., and the U.K. have grown somewhere between four and six times larger from 1980 to 2020. In spite of the GDP—generally considered a nation's annual "income"—growing so much, the economic outcomes have not been equally distributed. The rich are richer, the poor poorer, and the middle class in these and many other Western nations has shrunk significantly. Shockingly, life expectancy is declining in the U.S., considered the world's "greatest" economy. According to research by evolutionary biologist Peter Turchin, such gross income inequality historically has nurtured great social unrest, revolution, or civil war. In such contexts, people feel desperate and this emotional state is fertile ground for intergroup conflict, scapegoating, and snake-oil salesmen selling simple messages of hate as solutions to complex social problems.[45]

Intertwined with this social backdrop is also the history of 9/11, the so-called war on terror, and related propaganda. Media stereotypes of Muslim peoples as dangerous, backward, democracy-hating, fanatical, and violent have been prolific. With stories like that about a group of people who are religious, cultural, and ethnic outsiders, how could the emotional cores of the people of Hérouxville—like those of the rest of the Western world—not be overactive?

Similarly, the hateful anti-immigrant rhetoric of Donald Trump and his allies landed easily in the troubled ears and unsettled bodies that were already seeking answers to why America was no longer as "great" as it was in their memories. The feeling makes it easier to put Muslims and immigrants into the mental categories of "threat," "barbaric," "violent," and "criminal." Developing a town charter to protect your community or building a giant border wall with Mexico, therefore, could seem like rational choices, even when there's little evidence of any real threat.

And this is why emotional literacy skills are so important to issues of diversity and political differences. They allow us to discern what are real versus imaginary threats. If we do not develop these skills intentionally, we risk living our lives on autopilot, our choices and behaviors governed by unconscious habits and fears.[46] And when we're on autopilot, we may default to using the most readily available stereotypes, thereby living in a state of guardedness and suspicion. We tilt away rather than toward those who are different from ourselves.

Body Language: Our Early-Warning System

To state the obvious, emotions exist at both overt and covert levels. When we get angry, sad, or happy, the feeling has to break our particular personal threshold before we become *aware* of experiencing the emotion.[47] Before that threshold is reached, many of our feelings remain hidden within our unconscious. And that's a problem, because without our full awareness, those emotions influence our behavior, thoughts, and choices.

But there is a way to get a jump on what's happening to us internally: to notice body language and tone of voice, both in ourselves and in others. The feelings we don't express overtly are often conveyed through our bodies. For example, researchers have known for some time that people express their bias regarding racial others by sitting farther away from them, making less eye contact, and displaying increased facial muscle twitches. These signs indicate high levels of anxiety and nervousness.[48]

Unconscious body language is difficult to control. It may demonstrate our tendency to tilt away from (rather than toward) out-group members. Even actors—who are trained in the art of body language—are often unable to hide their racial bias.

A brilliant study led by Max Weisbuch from Tufts University in Massachusetts used popular TV shows to observe the body language of actors. The study found strong anti-Black bias, even though the Black and white characters in these programs were social and economic peers.[49] Ten-second video clips were created with audio removed and one character ingeniously cropped out so that their race was not apparent. Viewing the clips, impartial observers found that positive body language—such as smiling, nodding, and leaning in when talking—was far less common when white actors interacted with their Black rather than white counterparts.

Nalini Ambady, one of the coauthors of the study, bluntly stated in an interview that Black characters were "less liked non-verbally than white characters."[50] Such negative feelings portrayed by the unconscious interaction between actors—which is a form of anti-Black, pro-white bias—have a direct impact on the rest of us. In another phase of the study, the authors found that viewers were negatively affected by what they were viewing. Watching such subtly pro-white clips from TV shows (normally formatted with appropriate characters visible) resulted in higher pro-white scores on tests that measure unconscious bias. Putting this impact into context, according to the study, the eleven TV shows had an average weekly audience of nine million Americans each. This hints at the enormous impact of media alone in reinforcing existing racial bias in all of our lives.

So what's going on? Why all this body language bias?

It's a stretch to believe that, across eleven different TV shows, directors overtly and consistently gave their white actors directions to single out their Black peers for subtle negativity. That would be plain weird. But the researchers did indicate that they were uncertain whether the negative body language was scripted by directors, an innate reaction by white actors, or some combination of both.

For a number of years, my creative outlet was independent filmmaking and I'm aware that the actor's instrument is their body. Accessing unconscious reactions and emotions is the real craft behind the work. It wouldn't be difficult to argue that the study did reflect the unconscious pro-white bias that the actors held. And why shouldn't it?

Actors live in the same society as the rest of us. Their job is to express their unconscious feelings convincingly to create believable and real characters. As the next chapter will explore, we all possess unconscious bias. Anti-Black prejudice, to a greater or lesser degree, has been widely absorbed by North Americans. It would make sense, then, that actors who are trained to unleash their unconscious through body language would more readily reveal such bias. These performers, in essence, serve as cultural mirrors. They reflect back something unpretty that exists inside all of us.

To manage rather than be controlled by our feelings, then, we need to develop an early-warning system to the emotions bubbling below the surface of awareness. Self-awareness is the tool required for such advanced detection, the foundation upon which all other inner skills are built.

Inner Skill 1: Self-Awareness

According to Michael Inzlicht, a neuroscientist at the University of Toronto Scarborough, "there is substantial evidence that those with more executive control are able to regulate their prejudiced responses. People who are better able to focus their attention and manage their emotions tend to be people who are able to regulate their stereotyped associations."[51]

Executive control refers to the work of the prefrontal cortex, including planning, evaluating, thinking about ourselves, and impulse control. And executive control is premised on self-awareness, the starting point for inner-skill development.[52]

Self-awareness starts with attentiveness to our own emotions and needs. It includes knowing our strengths and weaknesses as well as having a strong sense of our worth and capabilities. It is the ability to self-reflect, follow our instincts and gut reactions, and be aware of the impact we have on others and the world around us (and of their impact on us).[53]

Even with a good handle on our conscious selves, it's the elusive unconscious parts that live in the shadows of awareness. Learning to direct our focused attention to the internal workings of our mind is critical to living a life where our actions and choices are aligned with our values. Especially regarding issues of racial difference and inclusion.

Researcher and psychiatrist Dan Siegel argues that developing such inner knowledge—what he calls *mindsight*—helps us "name and tame" our emotions, so that we know how and when to constructively process and express them.[54] It also helps us counter the sweeping emotional charges that underlie intergroup interactions, especially when there is competition or conflict. For example, such insights might have been useful to the leadership at Hérouxville as they began developing an unnecessarily inflammatory town charter in reaction to a perceived—but nonexistent—threat of outsiders.

The most extensive process for developing self-awareness that I'm aware of also happens to be the second inner skill, *mindfulness meditation*. This technique offers simple exercises for the brain that include attention to breathing, body sensations, and relaxation.[55]

Inner Skill 2: Mindfulness Meditation

Prejudice and stereotypes, as we have seen, are simply neural habits. As such, they are subject to neuroplasticity: they are flexible and can be altered through conscious attention. Mindfulness meditation has been shown to help change negative habits of the mind. It is the tried-and-true method of over two millennia for improving our focused concentration.[56] It's a specific form of attention that emphasizes our here-and-now experience. Mindfulness meditation is about being aware of what is happening in both the mind and the body, without reacting or judging.

This Eastern contemplative tradition has spread across the Western world over the last several decades. It has been modified for use in a variety of nonreligious settings, including health care, personal growth, general stress relief, and leadership development.[57]

In his book *Mindsight: The New Science of Personal Transformation*, Dan Siegel discusses the many benefits of mindfulness meditation. It can enhance resilience—our ability to bounce back from hardships—helping us tilt toward rather than away from challenging situations and people.[58] Further, from a neuroscience perspective, studies on long-term meditators suggest that we can literally grow and thicken the fibers in our prefrontal cortex through mindfulness practices, thereby enhancing our cognitive and emotional capacities.[59]

There are many ways to learn more about mindfulness meditation. Resources by teachers such as Thich Nhat Hanh, Pema Chodron, and the Dalai Lama are readily available, and there are local practitioners in many small and large urban centers. The most rigorously tested technique I'm familiar with is the Mindfulness-Based Stress Reduction (MBSR) program, developed by Jon Kabat-Zinn at the University of Massachusetts Medical School. Dr. Kabat-Zinn has written a number of books on the topic and has helped spread mindfulness across the health care sector.[60]

OTHER STRATEGIES FOR DEVELOPING SELF-AWARENESS

Many strategies besides meditation can also help us develop self-awareness broadly and recognition of our own psychological patterns specifically. Although beyond the scope of this book, the following may offer some starting points:

- Notice your own body language and tone of voice at regular intervals during the day. Track especially what happens when you get anxious, uncertain, or upset (clenched fists, irregular breathing, obsessive behaviors or thoughts).
- Take three to five opportunities daily to notice the shifts in your emotional state. Develop a broader palette of words to describe primary feelings (anger, joy, fear) as well as secondary ones (envy, contentment, nervousness).
- Recognize what issues, people, and situations emotionally trigger you into a state of fight-flight-freeze, especially regarding issues of racial difference. Everyone goes somewhere emotionally off-center when triggered—where do you go?
- Keep track of daily events in a journal. Review them over time to identify your patterns of choices, reactions, and behaviors.
- Get feedback from trusted others. Ask them specifically to help you consider perspectives that may be in your personal blind spot.

Questions to spark racial pattern recognition regarding our identities and differences can also help enhance our self-awareness. The following sample questions are adapted from cultural proficiency educator Randall Lindsey and his colleagues:[61]

- To what social identity groups (including race, gender, class, sexual orientation, and ability) do I belong?
- How are institutions and organizations in this country influenced by the dominant ethnoracial culture?
- How has my race and identity helped or hindered my progress in society, in small or big ways?
- How does race and social identity help or hinder people in my organization?
- How does my perceived status based on social identity in an organization (or society at large) affect my behavior and motivation to achieve? In general, how might perceived status affect behavior and motivation to achieve?

Besides these reflection questions, what new learning from this chapter helps expand your racial pattern recognition tool kit? What are systemic patterns due to socialization in mainstream culture and institutions? What new learning may also be useful in developing your psychological pattern recognition skills, those rooted in the neurobiology of humans?

IT'S NOT EASY to confront parts of ourselves that we are less aware of or that are contradictory to our espoused values. It can fuel painful emotions such as guilt, shame, anger, or defensiveness. This is where Deep Diversity's compassionate approach becomes important. Self-compassion helps us observe ourselves with curiosity rather than judgment. It's the salve to lessen the painful sting of our mistakes so we don't beat ourselves up. Yet it still holds us accountable. Compassion is essential; without it, we may not be able to focus our attention long enough to learn about and unlearn some bad habits about relating to others.

Finally, the key to developing any skill is practice and repetition. Although this may seem obvious, it's still worth mentioning.

Persevering is the hardest part of any habit breaking and forming process. If you're like me, it's an imperfect series of forward and backward steps. So, practice noticing your body language and breathing, even if there's a stretch of days in which you don't. Continue to ask yourself about the impact of your social identity on each situation, even if it's an afterthought. Practice. Rinse. Repeat. Do this until it becomes automatic.

Acknowledging this challenge from the onset may help us push through periods of inconsistency without getting demoralized. In this case, "fake it till you make it" is a completely acceptable principle. It may also be the most realistic path of learning for most of us.

Bias: Prejudice without Awareness

A Flutter in the Chest

I had just finished a meeting and was walking toward the subway. Although it was a little cool, I felt overdressed in my suit and overcoat. My briefcase weighed in my hand. I had recently left my public school teaching job to work independently, and this had been an important meeting that could develop into a longer-term diversity contract.

As I neared the subway entrance, somewhat lost in thought, a young woman approached me. "Do you know where the government offices are?" she demanded, in a flustered tone.

I didn't know but asked if she had an address. She rummaged around in her knapsack looking for it. She had lost her wallet the previous day, she fumed, and hated having to replace all of her identification, health, and credit cards.

When she pulled out a paper with the address, I didn't know what direction to point her in. The city block and the building in front of us were so big that it was hard to tell whether the numbers were increasing or decreasing. But I offered to help her locate the office.

I did this for two reasons. I had some time on my hands, and mobility was an issue for her—she used a wheelchair.

She readily accepted my offer and we proceeded down the street together. We exchanged names and started talking. We hit it off immediately; within minutes, we were chatting like old friends. Natasha was witty, sarcastic, self-deprecating, and even a little flirtatious.

After we crossed a large intersection, she paused, looked me up and down, and stated with the utmost confidence, "Shakil—I bet you I know what you do for a living."

I had two conflicting thoughts: *How presumptive,* and *Audacious—love it!* This was completely in line with her gregarious personality, and part of why I instantly liked her so much. I was up for her little game. In fact, I felt that I had an ace up my sleeve, especially with the conservative suit-and-tie look that I was sporting.

"Go ahead," I challenged. "Tell me what I do for a living."

Natasha scanned me again and pronounced, "You're either a psychologist, teacher, or social worker."

My jaw hit the ground. Besides the fact that I had just left teaching, if there were three professions that described the unusual job I'd created for myself, it would be those. I was flabbergasted. She looked on smugly as I affirmed that she had guessed right.

"Yeah," she continued, "from my experience, those are the type of people who stop to help strangers most often on a street... and on top of that, I'm a social worker!"

It was with the last part of her statement that I noticed a little flutter in my chest. This body twitch was subtle but I knew it meant something important, something I didn't want to acknowledge. It still evokes some shame to admit this, but I know exactly what that little flutter indicated. I did not expect her to say that she had a career as "elevated" and "respectable" as a social worker.

I expected less from her.

My surprise at her profession revealed something ugly in the depths of my unconscious: bias against someone in a wheelchair. Lower expectations of people with physical disabilities—this is called *ableism* and is something I teach people to recognize in my diversity and inclusion training. Am I not supposed to be the expert on these issues?

This form of unconscious prejudice—known in scientific terms as *implicit bias*—cares not if we are laypeople or experts. We all have implicit biases; it's a part of being human. Such prejudice, hidden in the realm of the unconscious, influences our behavior.

There is hope, though. Conscious effort can be applied to manage and reduce this masked form of bigotry that perpetuates feelings of "Us" and "Them."

Implicit Bias: Prejudice below the Radar

> *Deep within our subconscious, all of us harbor biases that we consciously abhor. And the worst part is: we act on them.* SIRI CARPENTER, social psychologist[1]

The term *implicit bias* requires some work to understand fully. Implicit refers to "mental associations that are so well-established as to operate without awareness, without intention, and without control."[2] In contrast, *explicit* feelings can be consciously detected, expressed directly, and publicly reported.[3] In the words of Mahzarin Banaji of Harvard University, the researcher who coined the term "implicit bias":

> When we speak of implicit bias, we are talking about decisions that people make that are happening quite outside their conscious awareness but nevertheless have a systematic pattern to them. Those patterns tell us that they are using information about a person's group membership such as their ethnicity, gender, sexuality, religion, culture, or language—the list goes on.[4]

I'll simplify matters and define implicit bias as a hidden or unintentional preference for a particular group based on social identity such as race, gender, class, ability, or sexual orientation. It's a form of prejudice that is indirectly expressed, originating in the unconscious mind.

We hold beliefs about social groups at both detectable and undetectable levels of awareness. We commonly refer to these beliefs as *stereotypes*—generalizations about a group of people that can be based on a kernel of truth or an exaggerated reality, or can even be an outright lie, resulting in the conscious or unconscious categorization of each member of that group without regard for individual differences. It's seeing someone as a symbol of a group rather than as an individual.

Some stereotypes are obvious and overt (such as Latin lovers, or women as homemakers and men as moneymakers). Implicit stereotypes, however, are subtler. They may be revealed through circumstances, when beliefs held below the radar of awareness are suddenly confronted. That's what happened in my interaction with Natasha. The low expectations I held of someone who used a wheelchair were uncovered.

Bias Helps: It's a Necessary Brain Function

It may surprise many of us to realize that implicit bias and stereotypes grow out of normal and necessary functions of the human brain. They are critical to how we perceive, categorize, remember, and learn about the world around us.[5]

The brain is a network of a hundred billion neurons, single cells that make up the majority of the nervous system. Neurons send and receive electrochemical signals to each other, creating networks and pathways that act as a high-powered communication system that allows us to regulate our bodies internally and react to our environments externally.[6]

From birth, we start developing neural networks for everything from forming our first words and recognizing language to learning to crawl, walk, read, or ride a bike. Our brains are associative by design. We learn by making connections between things, clumping, grouping, and pairing concepts. For example, most of us in North America have learned to make the following associations:

- ice cream with sweet (rather than savory)
- hammer with nail (not with cotton ball)
- snake with dangerous (not with cuddly)

Each of these pairings is a specific neural network. The stronger the association between two ideas is, the more well-worn the neural path. In neuroscience terms, this is known as Hebb's Law: "Neurons that fire together, wire together."[7] The more regularly we think, feel, or do the same thing, the more ingrained the neural path. Over time this repeated association becomes a "default" setting, in essence, a *habit*.

The pairings or associations are learned from childhood, absorbed and shaped by the stated and unstated norms of our family, friends, and social and physical environment. This is a way in which we learn about our *culture*, a concept that encompasses any of the shared values, morals, behaviors, customs, and worldviews held by a large group of people.

Another way to think about biases is as filters that help us pay attention to relevant stimuli while ignoring less important things. One scientific estimate suggests we may be exposed to close to eleven million pieces of information per second.[8] Take, for example, walking down a busy street. There is so much sensory information coming at us at any moment—cars, people, colors, temperature, awareness of social norms, odors, sounds, obstacles, and the comfort (or discomfort) of the clothing we are wearing, to name just a few.

If we were required to process and filter all that sensory information using the conscious mind, we would be utterly overwhelmed (and, likely, incapacitated). Anyone who drives a car, for example, may recall the nerve-racking early stages of learning to navigate a two-ton mechanical beast through busy streets.

Fortunately, this is one of the gifts of the unconscious mind—the ultimate multitasker. It processes vast amounts of information at lightning-quick speed, categorizing, perceiving, prioritizing, remembering, and learning all at once. Comparatively, in conscious awareness, we can handle only about forty pieces of information at any one time. And we can only keep a handful of items in direct focus.[9]

Although the mechanism is not fully understood, Nobel laureate Daniel Kahneman and his collaborator, Amos Tversky, provided evidence that humans use *heuristics*—mental shortcuts to deal with the vast ocean of information to which we are always exposed—rather than rational thought.[10] It is hypothesized that these neural codes are the foundation of our pattern recognition skills, enabling quick, efficient judgments that assisted our ancestors to survive through the ability to rapidly distinguish poisonous from edible plants, predators from prey, or friend from foe.

Such neurological shortcuts allow humans to recognize repeated events and emerging patterns, the foundation to learning and development in all spheres of life. They act as filters to keep out extraneous information and help us focus on what's really important. Such pattern recognition is part of virtually everything we do, whether it is to correctly identify if a person speaks with an unfamiliar accent; a math equation is properly solved; a word is misspelled; music is joyous or sad sounding; or pajamas are most appropriate for bedroom or boardroom. Heuristics are also believed to be the mechanism behind intuition or "gut" knowing, sometimes described as our sixth sense.[11]

The problem is that such mental shortcuts are imperfect and can lead to errors in judgment, because stereotypes and implicit biases are also forms of heuristics. That's when they hurt rather than help us.

Bias Hinders: It Distorts Our Worldview

> *As a result of... pre-established filters, we see things, hear things and interpret them differently than other people might. Or we might not even see them at all!* HOWARD ROSS, author and bias consultant[12]

Mental shortcuts about social groups create problems for us because important information is usually missing or edited out, impairing the accuracy of our pattern recognition system. Howard Ross writes, "Our perceptive lens enables us to see certain things and miss others, depending on the focus of our unconscious. It filters the evidence that we collect, generally supporting our already held points of view and disproving points of view with which we disagree."[13] He correctly suggests that bias creates literal—not just metaphorical—filters, causing us to fixate on some things while ignoring others.

For example, commonly held stereotypes in our culture incorrectly support the associations:

- woman with frail or weak (not with martial artist)
- Black man with criminal or athlete (not with engineer)
- immigrant with poor English (not with public relations expert)

Such neural associations follow the path of least resistance, quickly linking a group with a commonly held belief. Stereotypes, in essence, function as neurological imprints. Once the pathways are sufficiently reinforced and racial or gender patterns are formed, they are resistant to change.

Further, a quick and dirty heuristic in the form of stereotype or implicit bias about a social group can lead to misjudgments of people and situations. Mahzarin Banaji and R. Bhaskar describe memory and belief as generally operating outside awareness or conscious control, thereby threatening fair treatment and objectivity. They write: "When stereotypes are unconsciously activated and used, two direct challenges to the implementation of fairness are posed: (*a*) perceivers and targets are often unaware of the steady and continuous rendering of judgments and (*b*) judgments are based on beliefs about targets' social groups rather than on targets' actions."[14]

Much of this dynamic has to do with the rapid speed of the unconscious mind, which outpaces our ability to think about the situation. The pattern recognition system of our unconscious is "on" at all times. One of its jobs is to continually assess the world around us, seeking to make sense of what is happening in the external environment, on the lookout for patterns related to safety and threat. As a result, it continuously makes judgments on all things, including people. This can be beneficial unless our biases—based on race, gender, or identity—have been activated, when they become harmful.

Although both perceivers and targets may be aware of some judgments taking place, it is unlikely that either would be consciously aware of the degree of these judgments.

In addition, it's clear that biased judgments are based on beliefs about social groups rather than on their actual behaviors.

Implicit Association Test: Measuring Our Bias

Considerable research on implicit bias has emerged since the late 1990s from Project Implicit, which was cofounded by Mahzarin Banaji (Harvard University) and her colleagues Anthony Greenwald (University of Washington) and Brian Nosek (University of Virginia). This project helps expand our understanding of prejudice and discrimination beyond the overt, obvious forms manifested through racial epithets, images of swastikas, or ideologies of segregation. Project Implicit researchers created the Implicit Association Test (IAT), a free online assessment that asks participants to pair words with groups of people, measuring which associations most easily come to mind. Measured in milliseconds, the faster responses reflect stronger unconscious stereotypes and intuitions. In essence, these associations reveal dimensions of our pattern recognition system—our prejudice—and are often better predictors of many behaviors than our explicitly stated opinions.[15]

Various versions of the IAT have been available to the public online since 1998, and millions of people have participated in these tests. Hundreds of studies have also been conducted worldwide using the results.

The following are five key findings from Project Implicit:

- **We all have implicit biases.** Humans have unconscious preferences in everything from race, age, and gender to sexual orientation and class issues.[16] Such bias is described as "the well-learned and automatically activated associations between psychological qualities and social groups."[17] Family structures, peers, education, and other societal institutions reinforce cultural norms—many of which originate from arbitrary preferences. These norms benefit some groups while marginalizing others. As a result, we unconsciously prefer some groups to others.

- **We are unaware of our implicit biases.** Negative associations toward various social groups are harbored below the radar of consciousness. They exist in hidden zones shadowed from our awareness. People are therefore surprised to learn that they hold such biases.[18] Further, research demonstrates that many groups of people—including the researchers from Project Implicit themselves—can behave in contradiction to their stated beliefs. Bias becomes especially pronounced in decision-making situations where time is limited or stress is high. It is difficult to be aware of this implicit bias because it regularly outpaces our ability to think about it.[19]

- **People differ in levels of implicit bias.** In relationship to a particular social category, we can possess high, medium, or low/no levels of implicit bias. A person may demonstrate low bias regarding race and high bias in another category, such as sexual orientation. Various factors, including the influences of family, culture, and social context, influence the level of bias in an individual and in a category.[20]

- **Group power buffers or magnifies bias.** Although all of us possess unconscious prejudice, the negative effects of bias are linked to social power and group status in society. There are high-status or dominant groups (whites, men, heterosexuals, educated, rich, able-bodied) and low-status groups (Black, Indigenous, people of color, women, LGBTQ2S+, working class, people with disabilities). High-status groups demonstrate greater favoritism toward themselves and higher implicit prejudice toward nondominant groups. In fact, there's a tendency for low-status groups to also unconsciously favor dominant groups. Overall, nondominant groups predominantly feel the impact of bias, whereas dominant groups are buffered by greater social capital.[21] (We will return to the theme of power in Chapters 5 and 6.)

- **Minority groups internalize negative bias.** Unconscious prejudice can erode both trust in and desire to be with our own ethnocultural or racial groups, especially if our identities are non-dominant and lower status. Patterns in studies consistently show that both high- and low-rank groups hold more implicit prejudice toward minority group members. For example, one sampling of 260,000 race IATs conducted over a nineteen-month period demonstrated that it was not only whites who demonstrated a strong pro-white bias. Latinxs and Asians exhibited equally strong pro-white biases.[22] Other evidence suggests that even when members of a minority group explicitly state that they prefer people of their own race, their implicit choices and behaviors show the opposite—that they, in fact, favor the dominant group.[23]

Such research demonstrates that in many categories such implicit biases are better predictors of our underlying mindsets than what we explicitly say about ourselves, not because we are being deceitful but because our personal contradictions are frequently hidden from our individual selves. And this is where the IAT may be of support, as taking a test provides an opportunity to self-reflect about when our conscious beliefs about equality may be contradicted by our unconscious, impulsive reactions.

The IAT played a central role in bringing the implicit bias conversation into the mainstream in the 2010s, helping society reflect on judgments and behaviors related to a range of issues from political attitudes, voting, academic achievement scores, consumer preferences, social evaluation, hiring decisions, educational outcomes, and criminal justice reform.[24]

Having said this, current research also concludes that while implicit bias itself is an unquestioningly valid concept in helping understand human behavior, the test has limitations. First, the IAT does not accurately predict behavior, as this was not something the test was designed to do. A low or high test result on racial bias, for example, is not an accurate indicator of what a person may actually do in a real-life situation. A 2018 meta-study of over two hundred studies of the IAT found that the results have been inconsistent: in some cases the IAT strongly predicted behavior while in other cases it did not.[25] Second, one's IAT measure may vary depending on a variety of factors including fatigue, hunger, or levels of distraction. The researchers behind the IAT have always dissuaded people from using it as an assessment tool to select candidates for jobs. It is generally accepted that the IAT is most accurate when looking at aggregate numbers—meaning, it reflects the bias present across large populations. And, as I said previously, it is very helpful for reflection, offering an accessible, judgment-free entry point for individuals and organizations to explore the topic of systemic discrimination and its real-life impacts.

Beyond the Lab: The Real-World Effects of Bias

To make the implicit bias results more connected to the real world, consider examples from a few social domains: the workplace, health care, policing, and education.

WORKPLACE RECRUITMENT: WHAT'S IN A NAME?

We were discussing the importance of pronouncing names correctly during a Deep Diversity workshop when a participant shared that their partner, José, had been struggling to get a job interview in spite of high qualifications in the finance sector. Then a small change was made. José changed the first name on his resumé to "Joe" and was surprised—but not entirely—that he started receiving numerous calls for interviews, in spite of everything else on the resumé remaining the same.

This variation of pro-white bias was similar to the optometrist story I shared in Chapter 1 and is well established in labor market research. A 2016 study demonstrated the effect of Black or Asian candidates "whitening" their resumés by anglicizing their names, removing references to Black or Asian student clubs or associations, and including recreational activities popular among whites like hiking, kayaking, or snowboarding. Of the 1,600 resumés sent to entry-level positions across sixteen U.S. cities, 25 percent of Black candidates who whitened their resumés received callbacks for interviews, compared to 10 percent who left ethnic details in. Among Asians, the results were similar with 21 percent callbacks for whitened resumés versus 11.5 percent for unaltered ones.

Previous studies in both Canada and the U.S. have shown comparable results. English-sounding names like Greg Johnson and Emily Brown had a 40 percent higher callback rate for interviews in the Greater Toronto Area compared to resumés with the same education and job experience but with "foreign-sounding" names like May Kumar, Dong Liu, and Fatima Sheikh.[26]

Such results can occur even when companies are actively looking for diverse candidates. In the United States, one study looked at what happened when people with white-sounding names versus Black-sounding names (like Tyrone, Leticia, and Jamal) applied to companies that indicated they were "hungry" for qualified racial minority candidates.[27] Despite those companies' stated preferences, the resumés of white-sounding names still triggered 50 percent more callbacks than resumés with Black-sounding names. Even more shocking, lower-skilled white candidates received considerably more interview callbacks than higher-skilled Black candidates.

A 2018 meta-analysis of twenty-four studies conducted over a twenty-five-year period starting in 1989 indicates that little has changed with regard to discrimination in hiring practices for minoritized groups based on race and ethnicity.[28] Names and identities matter—white people, on average, receive 36 percent more callbacks for interviews, whether determined through fictional resumé audits such as the previously cited studies or whether the research was conducted in person with racially dissimilar but otherwise comparable pairs of trained candidates who apply for jobs in person.

HEALTH CARE: STEREOTYPE-CENTERED CARE

Name biases also appear in the context of health care. A 2019 study found that gender and racial biases were at play during professional medical conferences. Female speakers were less likely to be addressed with "doctor" and more likely to be addressed by first name only as compared to when the speaker was male. Speakers who were Black regardless of gender identity were less likely than their white counterparts to receive a professional address. Such informality suggests the person being referred to is of lower status, reflecting both gender and racial biases.[29]

Identity drives similar results in those who receive health care as demonstrated vividly by a study that involved 220 internal medicine and emergency room doctors from Boston, Massachusetts, and Atlanta, Georgia. They were required to assess a hypothetical male patient showing symptoms of a heart attack.[30] All of the physicians were presented with the same patient profile, the only difference being the accompanying photos: they showed either a Black man or a white man, both middle-aged. The doctors were also given a race-based IAT as well as a questionnaire that determined their explicit attitudes toward race. The researchers wanted to know whether race played any part in the doctors' assessment of the patients, specifically as to whether to provide thrombolysis, an anti-clotting therapy for heart attacks.

Although the physicians reported no explicit preference for Black or white patients, the IAT (specifically, the Black-white race test) demonstrated something different. The doctors—like any random group—fell into three distinct categories of unconscious racial bias: low, moderate, and high. The doctors who tested with high levels of anti-Black bias were much less likely to give the life-saving anti-clotting drugs—the correct procedure based on the patient symptoms—to Black patients.

Even with the limitations of the IAT, results of this study help explain what is widely known in medical research, which is that whites are twice as likely as Blacks to receive thrombolytic therapy for heart attacks. This means that Black people are *under*prescribed drugs that could save their lives.[31] In my opinion, this is the real power of the IAT, as it can serve as a key tool in the decision makers' toolbox to tackle the problem of racism within the medical profession in a way we could not previously: by examining the potential unconscious decision-making that may be a factor in underserving Black, Indigenous, and people of color communities.

Furthermore, bias appears to play a role in a cycle linking poor communication with negative health outcomes. As described in a 2017 paper for the American Psychological Association by Tori DeAngelis, "overall, providers high in implicit bias were less supportive of, and spent less time with, their patients than providers low in implicit bias. And Black patients picked up on those attitudes: They viewed high-implicit bias physicians as less patient-centered than physicians low in this bias. The patients also had more difficulty remembering what their physicians told them, had less confidence in their treatment plans, and thought it would be more difficult to follow recommended treatments."[32]

And sometimes racial bias prevents patients from receiving any help at all, even within a universal health care system like Canada's. In 2017, award-nominated rapper John River, who happened to be young and Black, was experiencing intense pain, shortness of breath, and sensitivity to light the day after a spinal-tap procedure in Toronto. When he returned to the hospital, instead of receiving support, he was accused of being a drug dealer or an addict, or of not being educated enough to know his symptoms were imaginary, despite the fact that his symptoms were known potential side effects of his procedure. It took him another sixty days of intensely debilitating pain, humiliation, and visits to five hospitals before he was diagnosed with a spinal fluid leak and received the appropriate care: an epidural blood patch, which he was supposed to have received within forty-eight hours of his original procedure.[33]

Such problematic issues and biased practices are, unfortunately, common in the health care context.[34] We've known this ever since the groundbreaking 2002 study called "Unequal Treatment" by the Institute of Medicine in Washington, D.C., which found significant evidence of discrimination playing a role in the differential treatment of racial minority and Indigenous patients compared to white patients—ranging from basic physical exams and history-taking to referrals for advanced treatment required for diseases like cancer to higher neglect and mortality rates.[35]

POLICING BIAS: MISTAKING WALLETS FOR WEAPONS

Studies have also shown that individuals with higher levels of anti-Black bias are much more likely to mistake day-to-day items such as a wallet or cell phone for a weapon when it is in the hands of a Black person rather than a white one.[36] Known as *weapons bias*, this is another focus of implicit race bias, especially in the context of policing. Banaji and law professor Curtis Hardin reviewed over two dozen experiments on weapons bias and found the results consistent.[37]

To break down the basic experiment, participants were shown a series of images and instructed to quickly "shoot" if the person is armed and "don't shoot" if the person is unarmed. The results were predictable: participants were more likely to confuse harmless things like a camera or soda can for a weapon when held by a Black person and therefore "shoot" these unarmed people. Participants were more accurate when viewing images of white people in the same circumstances, resulting in significantly fewer unarmed whites being shot. These results are, unfortunately, similar for not just whites taking such tests but also for Asians, Hispanics, and even Blacks. This demonstrates how minority groups who are the targets of such bias also internalize negative beliefs. Further, the results are similar among police officers. Banaji and Hardin bluntly state, "Such findings have important implications for police officers given the broader finding that police consistently use greater lethal and non-lethal force against non-white suspects than white suspects."

BIAS IN LEARNING: STEREOTYPE LIFT AND THREAT

A growing body of research demonstrates that implicit associations in the form of stereotypes are powerful enough to create self-fulfilling consequences in learning and education.[38] Individuals can live up or down to beliefs about their group, known as stereotype *lift* (positive generalization) or stereotype *threat* (negative generalization), respectively. Here are some examples.

In a groundbreaking study conducted at Stanford University in the mid-1990s by Claude Steele and Joshua Aronson, high-achieving African American students took the Graduate Record Examinations

(GRE). The control group took the GRE under neutral conditions, while for the second test group, it was implied that the GRE was an intelligence test. Test group 2 did considerably worse than the control group. Researchers concluded that the stereotype threat—based on the incorrect belief that Blacks are psychologically inferior to other groups—was so powerful unconsciously that it caused the lower scores.[39]

Another study went a couple of steps further, focusing on female East Asian college students and their relationship to stereotypes regarding mathematics. The researcher cleverly introduced the generalization that "women are worse at math than men" through a subtle prime—the participants were simply asked to complete a pre-assessment that asked questions that would get them to think about their gender identity, such as if they preferred living in a mixed-gender or single-sex residence. The test scores for these East Asian women were significantly lower than for the control group of East Asian women who were not asked to think about their gender in a pre-assessment. However, when the stereotype that "East Asians are generally better at math than other groups" was activated—through questions that asked about their family and ethnicity—the scores were higher than in the control group, with gender "threat" disappearing. These results demonstrate not just the power of stereotype threat and lift but also how easily our unconscious minds respond to subtle suggestions.[40]

To explore how much of this effect exists outside the classroom in a broader context of learning, another study focused on white golfers who were told they were being compared to Black golfers. (In reality, there was no Black group.) When told that the research was focused on which racial group possessed "natural athletic ability," the white golfers did worse than the control group who were given no generalization (demonstrating stereotype threat). When the white participants were told that the study was to determine which racial group had the greater strategy and golf "intelligence," their scores improved significantly (demonstrating stereotype lift). This study verified the continued existence of the historically incorrect belief that whites are intellectually superior, while Blacks are naturally athletic. Claude Steele described the invisible power of these implicit stereotypes as a "phantom" that permeates the air of society and haunts our social interactions.[41]

Overall, such studies provide ample evidence that unconscious generalizations are pervasive and have impacts both hidden and overt in many aspects of our lives. Negative stereotypes can be painful, reduce performance, and limit one's sense of self.[42] However, it's worth saying that all stereotypes are harmful whether positive or negative, as they box people into a small reality of who they are "supposed to be" according to others, rather than who they actually are according to themselves. Although stereotype lift is used in the context of studies to understand human behavior, it should never be used in an attempt to increase performance in a workplace or in society broadly. On the contrary, it would be highly unethical to do so.

Regardless of lift or threat, stereotypes do more harm than good and should be avoided.

Bias: Nature or Nurture?

An ongoing debate is whether prejudice is ingrained in our biology or the result of our culture and society. Is Us versus Them a matter of nature or nurture?

The research suggests that it's a bit of both, but in surprising ways that many of us may not have considered. In short, we are born with the bias hardware, while society provides us with the software.[43]

There is significant evidence that implicit bias has a neurological base, with the roots of Us and Them firmly set in unconscious processing. Various studies show that when faces of people of other races are flashed so quickly on a screen in front of us that conscious awareness is impossible, the threat response of the amygdala is still activated.[44] Other research suggests that our attention—indicated by electrical activity in the brain—commonly prioritizes those of our own race.[45] And our behavior regulation centers become more active when we are being politically correct and fearful of making mistakes.

These results are seen as early as the first year of our lives. Studies have demonstrated that even nine-month-old infants respond more quickly to people of the race of their caregivers than to others. Newborns are able to match emotional sounds with facial expressions faster and differentiate more rapidly between faces of their own race.[46] Before five months of age, however, babies respond to different groups in a more or less similar manner.[47] Other preferences are also established within this short time period, including greater responsiveness to the gender, language, and accents of the primary caregivers.[48] So, although we may be born prejudice-free, we learn bias at a very early stage. This suggests that we come pre-equipped with some neural mechanisms that seek to define and identify "our people."

Why? Again, some researchers point to evolution to help us understand our reaction to group differences.[49] It is suggested that our cave-dwelling ancestors needed to know instantly who was part of the group and who was not. To mistake a foe for a friend may have cost them their lives. It would have been an advantage to be neurologically wired with an "alliance detection system" to instantly recognize a tribe member and react swiftly with fight or flight.[50] From this perspective, we are the progeny of those who were hypervigilant and survived. Our ancestors are those who preferred to err on the side of caution and mistake (and mistreat) a few friends as enemies, rather than risk being those who were mistreated.

It appears that part of our biology causes us to tilt away from—rather than toward—those we perceive as different than ourselves. I remember being shocked when I first learned this. It went against my long-held belief that socialization was the only force to consider when it came to issues like racism or sexism. But the hardware we come with seems to matter in ways I had never contemplated.

Now, as the research on babies reveals, who gets defined as "us" is learned from exposure and experience from all our spheres of contact. Our software is determined by our culture and context. In the upcoming chapters on identity and power, we'll explore in greater depth the roles of socialization and systemic discrimination in establishing social norms. But suffice it to say that both nature and nurture have a significant place in forming unconscious bias, stereotypes, and racial intuition.[51]

It's important to state, though, that although our hardware may not be changeable, there's evidence that our software is.

Bias Reduction Strategies

In 2007, the results of my first IAT on race revealed that I had a moderate preference for white people over Black. I was embarrassed and devastated. As someone who works in the field of diversity and prejudice reduction, this was a significant result that led to some profound reflection. Part of me wondered if I should quit my day job. I sat with the results for many months, thinking about their implications for my work.

My results, however, did make some kind of sense. On a personal level, they fit my earlier life story. I am of South Asian ethnicity, and I grew up in small-town Canada. Part of me desired to be white. I worked hard to assimilate and "fit in." On a broader, societal level, pro-white preference is part of the collective North American story. As Project Implicit research demonstrates, the majority of both white and nonwhite people have pro-white bias. We've all drunk from the same cultural punch bowl, and our tongues are stained similar colors.

So, what to do? I started by doing what I knew best—I asked questions, read, and sought out new research while also revisiting studies addressing implicit bias. I found evidence that although implicit bias is consistent over time and rooted in unconscious processes, it is not completely fixed. That means we can change it. I was struck by the simplicity of one particular approach developed by researcher Brandon Stewart. He instructed study participants to use a counterstereotype—specifically, the word "safe"—whenever they encountered a Black person, and found a reduction in anti-Black prejudice.[52]

Given what I was learning about how the brain works, Stewart's strategy made sense. If stereotypes are simply an overused neural pathway—with the grossly incorrect connection between Black people and danger being particularly entrenched in our minds—then just telling myself to not make the association would likely fail. Stewart's strategy suggested that I needed to update my pattern recognition system by creating a new association between Black people and positive qualities.

Before I go any further, I want to acknowledge this may be painful or triggering for many to read, especially people who identify as Black. I find it beyond deplorable that, in this day and age, other humans would be described as "unsafe," as somehow "less than," because of an arbitrary measure such as skin color or ethnicity.

Yet my commitment to tackling my own anti-Black bias propelled me into self-reflection and then action. Over the next few years, I began a simple experiment with myself, usually when I was riding the subway.

Public transit allows natural time, space, and social permission to people-watch (without being creepy!). Leveraging the opportunity, I invisibly worked to upgrade my faulty pattern recognition system. Whenever I noticed Black people in the subway car, I would close my eyes and intentionally make a positive association: *kind, generous, philosophical, hardworking, engineer*. I repeated the words to myself several times while trying to picture the person's face with my eyes closed. Tackling my own anti-Black bias became a mini-habit, a regular way to pass a minute or three of my time.

In November 2012, five years after I took my first IAT, I repeated the process. This time, the results indicated that I had "little to no automatic preference between white people and Black people," the lowest level of the IAT. Although it's far from a scientific conclusion, I credit Stewart's research for helping me actively counter my anti-Black bias at the neurological level.

While I adopted this method, I was also conscious of a few things. First, that I was attempting to broaden the number of overall possible categories of Black people in my mind, with an emphasis on increasing my list of positive, nonstereotypical associations. (There are also positive stereotypes of Blacks: athletic, cool, good dancers, and so on.)

Second, that Black people are just another human group and so should not be romanticized as possessing only "good" qualities. (I did not, however, actively work to increase my negative list of qualities—society has done that adequately for us all.) But the awareness is important that within all our groups—within all individuals, really—exist the range and potential for a multitude of positive and negative human qualities.

And lastly, the criticism may also be leveled that to "gaze" at another person for our own learning in this manner is distasteful, perhaps even dehumanizing. And there is validity to such a critique. For this I am sorry. Yet, I'm aware that our pattern recognition system is "on" all the time regardless. Recall that our unconscious mind absorbs information from our surroundings, tracking information that reinforces bias and stereotypes. The question is: Do we want to do it unconsciously with a negative collective outcome, or update our pattern recognition software for a mutually beneficial outcome? It's a less-than-ideal choice, but choose we must. By guiding our conscious attention, we may be able to undo the unconscious habits of mind that hinder fairness between individuals and groups.

As I was writing this revised edition of *Deep Diversity*, I took the IAT again at the end of December 2020 and this time the results indicated I had a "slight automatic preference" for white over Black people. As disappointing as this was, it too made some amount of sense, if we return to some basic neuroscience. The brain has a "use it or lose it" rule—unused neurons are pruned away for the sake of neural efficiency.[53] The less we practice, the less efficient we are at the task at hand regardless of whether it's multiplication tables, playing guitar, or tackling our bias. I, like most other North Americans, spent the vast majority of 2020 at home due to the COVID-19 pandemic, only interacting with nonfamily members through online video platforms like Zoom. I was rarely on the subway or public transit and therefore had little real practice time to tackle my personal anti-Black bias patterns. As a result, although I was still improved from my original test in 2007, I was not as good as where I was in 2012 when I was in the thick of my regular inner practice.

This inner practice, known as counter-stereotypes, is one of many promising prejudice reduction strategies that researchers have shown can reduce bias.

Let's look at seven strategies that may help minimize such prejudice by upgrading our pattern recognition skills:

1. **Role models:** Various studies have shown that seeing people representing groups targeted by negative stereotypes in a positive light reduces bias. For example, exposure to strong, positive role models—whether Barack Obama and Oprah Winfrey in the U.S., Canadians Buffy Sainte-Marie and Jagmeet Singh, or Idris Elba and Gurinder Chadha in the U.K.—helps reduce bias, as does reading about the historical contributions made to civilization by Muslim-Arab communities.[54] Another study found that students in prejudice reduction classes taught by Black professors showed a greater decrease in both implicit and explicit bias at the end of term than those in a similar class taught by white professors.[55] (Similarly, female engineering students had more positive implicit attitudes toward math when taught by female professors than when taught by male teachers.)

2. **Inner motivation:** Research has found that people with an inner drive to be nonprejudiced are less biased. Bias researcher Michael Inzlicht has found that those who are intrinsically motivated (that is, who believe that both they and society are better off with less prejudice) tend to be more successful in bias reduction efforts than if the motivation is externally driven (they are pressured to comply).[56] Studies have also found that those with strong logic skills and willpower are able to better achieve their prejudice reduction goals, because they are able to notice their preferences and curb their judgments about others.[57]

3 **Noticing personal contradictions:** All of us display inconsistencies, to a greater or lesser degree, between our stated beliefs and how we act. Studies show that people who are able to detect the contradiction between their intentions and their actions are more successful in reducing bias.[58] Similar to cognitive-behavioral therapy as discussed in Chapter 1, such abilities help expose our pattern recognition system, at least momentarily. As stated at the start of the book, when we can identify the problem and name it, we can also tame it. Meditators are especially good at this. Their mindfulness training teaches them to observe their thoughts and feelings without judgment, a technique that tacitly familiarizes one with such discrepancies.[59]

4 **Intergroup contact and friendships:** Exposure to people different than ourselves helps curb the impacts of implicit prejudice.[60] Such bias was reduced among white college students if they were randomly assigned a Black roommate rather than a white one. Friendships between Muslims and Christians in Lebanon as well as between Blacks and whites in Chicago were found to improve implicit attitudes. In a U.K. study, implicit prejudice was found to be lower between white British and South Asian British children to the extent that they reported friendships. In an extension of this study, bias reduction occurred even among children who had no friendships outside their own racial group themselves but reported having friends who did.

5 **Counter-stereotype plans:** A substantial body of research demonstrates that using counter-stereotypes can help.[61] Brandon Stewart, as discussed above, instructed study participants to simply use words like "safe" whenever they encountered a Black face, and found a reduction in anti-Black prejudice. The key to this subtle reminder is that participants themselves create both an intention and a plan to tackle bias.[62] The stereotype, although not eliminated by this method, is diminished in impact because now another positive association is created in the subject's brain,

which may become more automatic over time. As mentioned, though, it is important to maintain an awareness that all groups have both positive and negative qualities. Attempting to paint a group as only positive is not only unrealistic but difficult for our conscious mind to accept.[63]

6 **Carrots can help:** Because a stereotype is a generalization, when we encounter a member of a racial group different than our own, there is a tendency for our brain to register that person as a symbol of the group, rather than as an individual. (This does not happen for members of our own group.)[64] Researchers have found that getting subjects to ask simple questions about vegetable preferences (as in, I wonder if this person likes carrots?) helps in bias reduction. It appears that the power of curiosity can help humanize others, so that we see them as unique individuals rather than as representatives of a group.[65]

7 **Education and training:** Various types of diversity training may have various results. But specific strategies, such as those listed here, have been shown to reduce bias in the long term. Patricia Devine and her colleagues at the University of Wisconsin–Madison worked with ninety-one non-Black students on the impersonal and implicit nature of bias and provided flexible strategies (including many of those listed here) to "break the prejudice habit." Over 90 percent of this group had a pro-white bias, as measured by the IAT, at the start of the study. After the education interventions, the results of the IAT showed a strong drop in implicit attitudes, results that lasted up to twelve weeks after the training.[66]

Inner Skill 3: Self-Regulation

All of these bias reduction strategies are premised on the first inner skill, self-awareness, and are greatly supported through practice of the second, mindfulness meditation. However, attunement to our feelings, needs, motivations, and behaviors is insufficient on its own to break our prejudice habits. *Self-regulation* is needed to manage the difficult feelings that emerge from uncovering bias.

We learn through our mistakes, which, unfortunately, can make us feel bad. This is not an easy process. According to Michael Inzlicht,

> Errors, fundamentally, are an adverse experience, an emotional experience. People don't like them, they don't want to make them. They sweat more, there's heart rate acceleration, pupil dilation, release of cortisol. It turns out the errors are really important in establishing self-control. People who are sensitive to the errors they make—who react and adapt to the mistakes they make—are really good at self-control and executive functions.[67]

Inzlicht suggests that if we listen to what our negative feelings have to tell us regarding our own prejudice and stereotypes, we can exert greater self-control.

Self-regulation (sometimes called self-management) is the third inner skill required for navigating emotionally loaded terrain and reducing bias.[68]

Self-regulation is the capacity to keep disruptive emotions and impulses under control, especially under stress. It's the ability to own our mistakes rather than avoiding them. Self-regulation includes having optimism, being able to experience positive emotions, and behaving with integrity and in socially desirable ways. It also includes knowing how and when to acknowledge and communicate important feelings, rather than simply repressing them. Self-regulation has been described as the inner skill that "frees us from being a prisoner of our feelings."[69]

In my encounter with Natasha, I could have ignored the flutter in my chest that indicated I had made a negative assumption about her skills and capacities. But I noticed it (self-awareness) and reflected on the experience. I tilted toward rather than away from my mistake, which required managing feelings about being "bad."

My inner critic can be harsh and loud, and my journey has been to learn from my errors without beating myself up. (Others, however, may be on the opposite end of the spectrum. They may need to learn to listen more closely to their inner critic and take responsibility.) Integrating compassion has been a key aspect of my own self-regulation practice. So understanding that much of what we struggle with is a normal part of the brain's architecture is part of updating my personal psychological pattern recognition system. Together, these insights allowed me to implement the subway strategy to reduce my anti-Black bias.

Whether we have named them or not, all of us have developed some practices to help us manage our emotions through difficult situations. We know that such strategies make us feel better, even if only temporarily. We may, for example, know that going for a walk helps us cool down after an argument. For some of us, venting to a confidant is useful, or writing in a journal. Constructive regulation strategies help us put things into perspective, problem-solve, and recover from or prevent difficult emotional situations. Some strategies, however—such as smoking, binging on food or alcohol, obsessively ruminating, or worrying—may be less constructive.

Many resources and supports are available on self-regulation, as well as a significant body of work. The following questions may serve as a useful starting point in expanding your psychological pattern recognition tool kit with regard to self-regulation:

- What are common things that regularly cause me emotional stress? Consider home, work, family, finances, and so on.
- What strategies (positive or negative) do I use to de-stress? Consider both low-intensity and high-intensity situations or conflict.
- What emotional situations trigger my fight-flight-freeze response, and how do I deal with these? Once off-balance, how do I recover?
- How regularly do difficult emotional circumstances appear in my life? What healthy prevention strategies do I implement?

Regarding self-regulation in the context of bias and racial differences, consider the following questions:

- What feelings emerge when I consider the possibility that I have biases and have likely acted on them, whether consciously or unconsciously?

- What was my response when I was corrected by someone for something I said about issues of difference (or when I was accused of being prejudiced)? What were my feelings? (If I haven't experienced such an accusation, how might I feel and react?)
- What issues about social identity (such as race, gender, or sexual orientation) am I most comfortable talking about? Least comfortable? Why are some easier than others?
- How often do I engage in conversations about race and difference that challenge and extend me beyond my comfort zone?
- What other self-regulation strategies might help me learn more about my own biases?

As we close this chapter, what new learnings about systemic discrimination issues help expand your racial pattern recognition system? What new ideas might you integrate to develop greater psychological pattern recognition?

The better we become at understanding social and individual stressors to help manage ourselves emotionally, the more effective we can be in learning and unlearning biases, and having critical conversations about our racial and social differences.

Although there may be bias toward dominant group members such as men, white people, heterosexuals, rich people, or able-bodied people, the negative impact of prejudice tends to affect non-dominant groups—including women, Black, Indigenous, and people of color, people with disabilities, poor people, and those who identify as LGBTQ2S+. As we will see in the next chapter, being part of the dominant group comes with social power, an invisible cultural momentum that supports us whether we are aware of it or not. This sets up the historical challenge facing society today, of good-hearted, well-intentioned people on both the giving and receiving end of prejudice struggling to work through our differences.

Identity: Belonging Drives Human Behavior

A Fish Tale

I live in a province that is blessed with freshwater lakes, both large and small. During one road trip through the countryside with a friend, I was struck by an unusual sight when we got out of the car for a short stretch break. On a pier extending into Lake Ontario were more than a dozen fishermen, all East Asian in ethnicity. Given that these small towns are predominantly white, the men on the pier really stood out. Amid snacks, hot drinks, and camaraderie, the anglers skillfully pulled some healthy-sized trout from the water. We started chatting with one of the guys and learned that they were from Toronto. This was their weekly adventure.

I left that interaction with a feeling of wonder. How much things had changed in the province since my childhood! I grew up in small towns like these, and—besides me and my family—it was a rarity to see racial minorities at all, let alone fishing. They were certainly not seen in such large numbers as I was witnessing with this crew of anglers. On top of that, these guys fished weekly! Amazing how demographics and lifestyles had shifted.

This gentle but important act of cultural integration filled me with hope and encouragement. Fish-loving cultures from the East meet angling towns from the West. What a great milestone!

That was about a year before I heard of the Asian angler controversy from this very same region.

What's that, you ask?

Well, imagine for a moment a man and his thirteen-year-old son fishing peacefully on the shore of a picturesque lake. The careful lesson of attaching a hook to the line. Bait wriggling in little hands. Fish tales from the past, transferring learning from elder to child. And some quiet moments in between.

Incidentally, the ethnicity of this pair is East Asian. And that's a factor as they are approached by two white men, locals who accost them. Threats and accusations are uttered at the father-son team, who are confused and fearful.

This was an example of a real event, a racially charged trend in small towns in my region in the mid-2000s. In another scenario like the one described, a thirteen-year-old child was grabbed and, in front of his shocked father, harshly pushed into the lake.[1] Over a period of a few years, victims experienced ethnic slurs and verbal harassment, as well as physical intimidation and violence. Property damage and physical assaults became commonplace. One violent incident put a young man into a coma, resulting in permanent brain damage.

Assaults on people of East Asian descent while fishing became a flash point attracting national media coverage, and a public human rights inquiry was launched to understand the nature and extent of the phenomenon.[2] The common factors were that the locations were small towns within a two-to-three-hour drive of Toronto, the hobby was fishing, the targets were always East Asians (and their friends), and the perpetrators were locals who were white.

There was much commentary on this issue, which tended to fall within three broad categories. Some say that this was an example of outright racism and the narrow-mindedness of small-town white people. Others say that East Asian Canadians were not following appropriate fishing rules and so brought this situation onto themselves. Lastly, some say this behavior stemmed from competition for dwindling resources, limited space, and declining fish stocks.[3]

While each of these elements has some relevance, I'd like to look at the phenomenon from a broader viewpoint. This situation exemplifies the third component of the unconscious mind—the human tendency to protect the interests of our own group and its members, regardless of whether we are white, Black, brown, or otherwise. This also includes our political affiliations as exemplified by the increasing divide between liberals and conservatives in the U.S. and events like Brexit in the U.K. And it identifies a key undercurrent responsible for the upward spike in hate crimes experienced by East Asian–looking people across North America and Europe during the coronavirus pandemic.

However, the simplicity of the intergroup dynamic between whites and Asian anglers can help shed light on these more complex patterns of political polarization, prejudice, and racism.

To undo the negative impacts of our group-ish nature, however, we first need to explore how deep its biological and cultural roots go.

The Neural Roots of Identity

> *We fear what our ancestral history has prepared us to fear: snakes, spiders and humans from outside our tribe.*
> DAVID MYERS, professor of psychology[4]

We have an inherent tendency, called risk intuition, to assess threats in our environment. According to David Myers of Hope College in Michigan, psychological science has identified an important factor that feeds our risk intuition: interacting with those not from our "tribe." It is theorized that risk intuition was a survival mechanism that allowed early humans to quickly distinguish a group member from an outsider—and a threat.

As we saw in the previous chapter, we learn to attune to the race of those closest to us starting in the first year of life. With repetition, neural networks form and create racial imprints, so to speak. Our brains don't have to work as hard when confronted with those most like us. This allows a sense of ease and comfort, because the job is being handled by our unconscious pattern processors.

This Us/Them formatting of our neural wiring happens quietly and automatically deep inside the brain, without needing overt encouragement. (Recall, from Chapter 1, the drinking-water study.)

Without our conscious awareness, our pattern recognition systems are constantly receiving and broadcasting emotional signals such as greater trust or empathy for strangers who share certain racial features with us.

If other races do not play a major part in our day-to-day existence, neural pathways countering these early imprints do not form. As discussed in the previous chapter, this is a variation of the brain's "use it or lose it" rule. The cumulative effect of genetics and socialization forms our unconscious racial habits and cultural norms.

RACE IS SCIENCE FICTION

Having said that, the concept of racial identity is rooted in society rather than in biology.

Race is a social construct; it's a concept created by people rather than something genetic in nature. Science long ago proved there is only one race—the human race—and that there is more genetic variance within a group than between groups[5]—meaning that a Black and a white person may be more similar genetically than two white people or two Black people.

Although there may be apparent differences between groups along the lines of skin color and other physical characteristics, extending this principle to imply that there are separate races is extremely dubious. When considering differences between groups of people, it is more accurate to discuss *ethnicity*—a shared identity based on cultural, linguistic, religious, historical, and/or racial factors (Pakistani, British, Somali, Jewish, and so on).

Yet, on a day-to-day basis, race matters, whether we agree with the concept or not. So it must be treated as though it were real. Author and academic Robert Jensen captures its complicated nature: "Race is a fiction we must never accept. Race is a fact we must never forget."[6]

Identity is even more complicated than race. Our sense of belonging to a group includes a broader range of socially important identities, including but not limited to gender, class, sexual orientation, age, religion, language group, and ability/disability. Our sense of identity can also change according to context and circumstance. In the workplace, interdepartmental rivalries can boost performance through micro-group identities (such as sales team A is outperforming sales team B). A national identity may suddenly become a point of group pride and celebration during global competitions such as the World Cup of soccer or the Olympics.

But when we get too invested in an identity, groups can also limit cooperation and positive outcomes. Sports fans can become unruly when their team loses or wins, and clash with opposing fans or police. Racial identities can also lead to violence and negative feelings, as in the anti-Asian bias demonstrated through the fishing controversy or the violence directed against East Asians during the COVID-19 pandemic across nations such as Canada, the U.S., and the U.K.[7] History teaches us that when intergroup tensions escalate to the extreme, the atrocities of genocide become possible, as demonstrated by Rwanda, Bosnia, Nazi Germany, and the historical attempts to exterminate Indigenous Peoples across the Americas.

Social scientists have studied the dynamics of groups, identity, and belonging for decades. Their findings have been synthesized into a framework called *social identity theory*. This theory is an evidence-based lens that highlights intergroup patterns, useful when attempting to understand the complicated dynamics of discrimination and racism in our organizations, as well as in society at large.

Identity Created by Culture

> *We generally are not highly aware of the rules of the game being played, but we behave as though there was a general agreement on the rules.* WILLIAM GUDYKUNST, professor of human communications studies[8]

My nephew Zephan lives around the corner from me. When he was about three years old, we developed a little ritual whenever I passed by his day care. If I happened to walk by when the children were out for recess, he would rush to the fence. I would crouch at his eye level and visit for a couple of minutes, me on one side of the fence and him on the other. Just before I'd leave, he would give me a happy little kiss through the bars. It truly made my day.

Then one time, I leaned toward the bar for our ritual kiss only to have Zephan pull back and gleefully yell, "*No* kissy!" I felt a surge of confusion and hurt. I also saw that a number of his little friends were standing behind him watching, including a couple of older boys. This was the last day he kissed me in front of his day care friends.

I later heard from my sister about Zephan's newfound conviction that "boys didn't kiss." For a short while, he even resisted kissing his mother. He was not taught this at home or by his day care providers. We suspected that he had picked it up in the social context of his day care peer group. Current research supports the suspicion about the formation of Zephan's gender identity, that children are influenced strongly by the gender "club" of their peers.[9] Although our peers are a significant part of the process, they aren't the only factor. It's deeper than just socialization; the process involves internal mechanisms that seek to define who and what we are.

Social identity theory explores these dynamics between individuals—their identity and behavior—in relationship to social groups. Pioneered by Henri Tajfel and John Turner, this body of work has developed through decades of research from around the world that have uncovered patterns that are common in human group behavior.[10]

It is widely accepted in research that belonging to groups is a key driver of human behavior, both on conscious and unconscious levels. Fundamentally, group membership helps us make sense of our place in the world. It removes ambiguity about others and ourselves.

An individual's self-concept—who we are—is derived from a combination of how we perceive ourselves in specific situations and our sense of belonging to various groups. These might be broad groups in society or specific ones in the workplace. And we frequently experience a tension between our need to belong and our desire to be seen as unique.

Our social identities are also heavily influenced by culture, all those ways of being, thinking, and acting that we learn from our environment. William Gudykunst, a scholar of multicultural communications, described culture also as a shared understanding of the "rules of the game" that emerges in an unstated manner and is specific to each context—the commonalities among relatively large groups of people.[11] We unconsciously learn these rules from a variety of channels including family, peers, media, schools, and society at large. As this happens, the role of groups in our lives becomes more entrenched in how we view ourselves, solidifying our sense of self. Hence, my nephew Zephan and his day care friends.

Belonging to social groups is an important part of our self-concept. Value and emotions are associated with such membership.[12] Our sense of gender, for example, begins very early. Many toddlers in the Western world, like my nephew, learn a number of gender rules by age three. For example, they learn that wearing a dress is "only for girls," as is the color pink. Our unconscious is easily trained by our environment, as it absorbs the lessons and stitches together mental shortcuts about what is and is not "normal." Our interactions with our peer groups, family, community, and media immerse us into these norms, patterns, and rules. They become an implicit part of who we are and how we make sense of others in the world around us.

If asked, for example, why dresses and pink are only for girls, our conscious mind will justify our behavior. It will find a reason even when there is little rationale, biological or functional, for such clothing or color segregation. Fundamentally, it boils down to a sense that the choices "feel" right to us. Although this heteronormative behavior is learned within many European-based nations—not all cultures share the aversion to males wearing a flowing, one-piece garment that extends down the legs—the feelings seem tied to neural architecture that demands that our tribal identities be defined.

Even within the same culture, what feels right changes over time. In the North American/European context over a hundred years ago, pink was considered the more decisive, strong color and given to boys, while blue was for girls. In fact, at the time, all children regardless of sex wore dresses till about age six because they were considered the most practical.[13]

And sometimes change occurs more rapidly than generations.

There is greater awareness today that gender identity goes beyond the binary of male-female to include transgender male, transgender female, and nonbinary people as well as those who choose to express their gender in a variety of ways. Although trans people still face considerable hardship and oppression, there is much more permission and support now at the start of the 2020s then even a decade before when society hardly recognized trans issues as a part of the human rights and dignity framework.

In-Groups and Out-Groups

Our social identity—including race, gender, class, and sexual orientation—is that part of our self-concept that derives from belonging to social groups. It's also tied to the identification of in-groups and out-groups, people whom we perceive to be similar or dissimilar to ourselves. Social identity theory can help us understand the complex intergroup dynamics and social alliances underpinning racial controversies such as the Asian angler issue in small-town Ontario.

IN-GROUPS: THOSE LIKE US

In-group patterning begins with people we perceive to be most like us. We care about our in-group and are taught to socialize with people from this group. We share the same references as to what is "normal" in communications and interactions. We unconsciously understand the unstated rules of the game. We are more willing to make sacrifices for our in-group and we do not always require equitable return from them. We tend toward generosity and a willingness to overlook mistakes made by members of our in-group.

Our alertness to our in-group patterns and members can be subtle yet powerful. For example, I find my radar sensitive to noticing strangers who share my South Asian ethnicity. My Jewish friends have commented on a similar awareness of those like themselves, as have friends who are gay. Each group playfully refers to such heightened awareness as "Jewdar" and "gaydar," respectively.

Besides such socially important identities, in-groups can be more fleeting and arbitrary. They can be based on the city you live in, the sports you play, or whether a researcher places you randomly on a blue or green team during an experiment. The patterns seem to be fairly consistent. There is a natural tendency to feel good about our in-groups, creating a positive sense of identity. This result is the formation of an *in-group bias*, which becomes especially pronounced if we belong to high-status groups in society.

The neurological underpinning of identity makes it easier to appreciate why we notice and prefer our in-group members, especially those tied to key social identities. Through a lifetime of being immersed in specific cultural patterns, the result is strong, unconscious neural imprints—we just intuitively sense what normal behavior is. Efficiency is a key design element of neural circuits. The less our conscious mind is needed, the less energy is required to handle the task. This same neural efficiency results in less awareness and therefore less control over some behaviors, actions, and choices.

OUT-GROUPS: THOSE NOT LIKE US

In contrast, out-group patterns begin with people we view as different than ourselves. Because we have not been taught to socialize with members of out-groups, we may feel uncomfortable in their presence. Out-group members are perceived as "all the same," whereas it's easier to recognize the nuances between individuals within our in-group.

Facial features are a classic example. If we haven't been socialized in an environment with East Asian people, it may be difficult for us to see the differences between those of Vietnamese, Chinese, or Japanese ethnicity, for example, let alone recognize the nuances within an ethnic group. If we're not thoughtful about learning and don't consciously leverage our neuroplasticity, our neural networks will not develop the necessary pattern awareness, making us susceptible to committing the *they-all-look-the-same* error.

The rules for social interactions and relationships may feel less clear with out-group members, creating conscious or unconscious anxiety. Our unconscious pattern recognition system, which seeks the familiar, struggles to comprehend out-group behaviors and environments. The conscious mind has to be more engaged, which requires greater energy and effort. If you've ever traveled to a new region, state, or nation where both language and customs are unfamiliar, you may have experienced this effect.

Because the unfamiliar challenges our pattern recognition system, there is a tendency for greater frustration and confusion in communications with out-group members, and our unconscious mind's desire to avoid out-groups is usually high. Everything is more work when engaging with out-group members because their social patterns and ways of being may be different than ours. This means we already feel under stress, so we are harsher in our judgments, we are less forgiving, and we expect more in return from out-groups.

Why? Because social norms are *emotionally* loaded and out-group behavior disrupts our sense of what's normal. A substantial body of research shows that most social norms only become visible when they are breached by out-group behavior. Out-groups behave in a way that inherently breaks accepted ways of doing things, and this experience is emotional, especially for the in-group. Take, for example, the simple handshake. A "good" handshake occurs with bodies about two to three feet apart; it requires a firm grip with two or three up-and-down motions as well as eye contact. I remember learning early in my socialization that a firm handshake was especially expected of men, whereas it was more acceptable for women to use something softer (though less so in a workplace context). I remember learning that a weak handshake by a man was not okay (out-group behavior); it was often given the sexist and heteronormative label "girlie." I suspect this may be why for some people, a firm handshake by a woman can be misinterpreted as aggressive.

Of course, the conventional handshake used in Canada, the U.S., and the U.K. is purely context-specific, tied to the white, Anglo-Saxon Protestant roots of these nations. Cultures across the world have traditions such as greeting with kisses on the cheek, like in Persian or Dutch cultures, or the lack of eye contact that signifies respect in many Asian traditions. In some conservative interpretations of Islam or Judaism, respect is also shown by not shaking hands or making physical contact with nonfamily members of the opposite sex. Post-9/11, the simple handshake became a flash point for anti-religious emotions. Many people, including politicians on both sides of the Atlantic, were offended when conservative Muslims or Jews refused to shake hands with members of the opposite sex.[14]

Through observational studies, theorists such as Harold Garfinkel have demonstrated that many social norms and practices (referred to as *background expectancies*) are taken for granted.[15] They become visible only when they're broken. People use background expectancies to participate in everyday interactions such as taking one's place in line, offering a certain amount of space during a conversation, or boarding a bus and sitting on a seat. People often perform these "mundane" tasks on autopilot, based on previous experiences and without question. We only become aware of them when norms are violated or "breached." As Garfinkel demonstrated, making the commonplace scenes visible reveals them. A person may feel uncomfortable if a stranger stands too close to them during a conversation. People will be irritated and maybe even angry if someone skips a line and walks to the front.[16]

In the context of 2020 COVID-19 public safety protocols to reduce virus transmission, there was an abrupt change in cultural habits in many nations as elbow bumps replaced handshakes and fist bumps as the new norm. However, not all new habits were easily adopted or welcome. There were many virus skeptics who became irate and aggressive with the new recommendations, especially requiring the wearing of a face mask.

So social norms that are usually invisible can, when they are changed or broken, become the source of controversy and emotions. Perhaps this, too, was an evolutionary adaptation that helped keep in-group members in line when the size of the community was extremely small. But it is less useful in a global village.

In the fishing story that began this chapter, the in-group/out-group dynamic is demonstrated by the violent reaction of the local white fishermen toward the Asian anglers. To break the patterns down further:

- The victims would identify with the *East Asian in-group*. From this perspective, the out-group consists of the local whites.
- The perpetrators are part of the *white in-group (local)*. From this perspective, the out-group includes the anglers of East Asian ethnicity.
- A few of the victims were also white, because they were friends of the East Asian Canadian anglers. They, however, might be described as part of the *white friends of East Asian anglers in-group*. Their out-group would also include the local whites, even though they all belong to the broader white in-group of society.
- Remarkably, both victims and perpetrators have a social identity in common: they would identify with the *angler in-group*. Anyone who does not identify with the hobby of fishing would be part of their collective out-group.

So, in-groups and out-groups are defined from the perspective of the individual (and group), and the concepts are very relative. We all simultaneously occupy a variety of in-groups and, therefore, out-groups. But there's one aspect of the Asian angler story that is crucial to discuss: power.

DOMINANT AND NONDOMINANT GROUPS: THE IMBALANCE OF POWER

In any story that involves more than one social group, the relative social power of groups involved affects the way it is told and understood. But not all in-groups hold equal status and power in society—a crucial point in understanding intergroup dynamics.

For historical socioeconomic and political reasons—conflict, colonization, cultural or religious norms, and so on—there are always high-status and low-status (or dominant and nondominant) groups in society.

So, we all belong to groups that are in-groups or out-groups relative to each other. It is when the power dynamic comes into play that an in-group becomes a dominant group. The following examples illustrate a number of higher-status groups that hold cultural and institutional power:[17]

- **Gender:** Men are the most powerful gender in-group in North America and in the world, dominating decision-making and leadership positions across societal institutions (government, big business, media, police, and so on). From this perspective, women and transgender people are nondominant or minority groups.

- **Race:** The members of the dominant racial in-group in North America share common white-European ethnicities, and control the levers of decision-making across the corporate, public, and not-for-profit sectors. Black, Indigenous, and people of color communities are proportionally underrepresented in the spheres of social power and are considered nondominant. (It wouldn't be difficult to argue that white people are the dominant racial group globally, in spite of their smaller numbers compared to their Indigenous and nonwhite counterparts.)

- **Disability:** Able-bodied people are the dominant group nationally and globally; societies are designed to meet their needs almost exclusively. Much less social consideration is given to people with disabilities, who occupy a nondominant or minority status.

- **Sexual orientation:** Heterosexuals form the most powerful in-group with regard to sexual orientation. People of diverse sexualities are part of the out-group.

- **Class:** The dominant in-group with regard to income includes middle-to-upper-class earners. Working-class and poor people are part of the economic out-group.

Taking power into account helps us more fully discuss the interplay of identity and racial dynamics in the fishing controversy.

How Identity Played Out in the Asian Anglers Controversy

The dominant in-group in the region where most of the Asian angler incidents took place—as it is in the nation—is white. In this area, Black, Indigenous, and people of color make up less than 4 percent of the population.[18] The targeted out-group was the anglers of East Asian ethnicity.

A human rights inquiry drew attention to a number of issues and contextual factors. A key driver was competition for limited space and resources. Over time, it had become more difficult to find spaces for fishing and other leisure activities on public places along waterways, such as piers, docks, and bridges.[19]

The white angler population of that region felt the competition for space more acutely than other groups. More people of *all* backgrounds started to find their way to fishing holes that had previously been frequented only by locals. However, the focus of the dominant in-group fell on the most obvious out-group—East Asian Canadians—rather than on other white people, outsiders who were less recognizable as such.

Earlier in this chapter, we noted that a group already under stress judges out-groups more harshly. In-group bias (those well-worn neural pathways) results in greater alertness and criticism of out-group members; the perception of their missteps is more pronounced. The inquiry found that locals denied race was a factor, instead focusing on the fishing practices of Asian Canadians and generalizing negatively. One local concluded that Asians had no respect for Canada after witnessing one occurrence of an Asian Canadian angler leaving garbage behind. Another local stood firm in his accusation that "Asians are raping our lakes," after witnessing a single incident of an Asian Canadian keeping undersized fish.

The inquiry also found that there was no evidence of any one group violating laws more than another. The psychology of the dynamics of groups is such that even if a white angler left garbage behind or poached fish illegally, there would be more leniency toward the in-group member, with no (or low) resulting consequences. There were no incidents of nonlocal whites being harassed or attacked for fishing in the same manner as the East Asian Canadian anglers (unless they were perceived as friends of this racial out-group).

The broader pattern is that nondominant groups not only are the targets of the dominant group's threats and assaults, but are also, according to the dominant group, responsible for any violence that happens to them. That dominant group bias moves easily from generalizations and incorrect beliefs to outright hostility toward out-group members is well illustrated in the case of contested space for anglers. But similar things happen all over the world. In this story, it's about the dominant white in-group literally attempting to push out the Asian Canadian out-group in the context of fishing in southern Ontario. In Myanmar (formerly Burma), it's the Buddhist majority brutally persecuting a Muslim minority who has lived there for generations.[20] In Pakistan, it's the Sunni Muslim in-group who are legally allowed to oppress the Ahmadi Muslim out-group.[21] Taken to the extreme, genocide is the ultimate demonstration of violent in-group bias. The 1994 Rwandan genocide was perpetrated by the Hutu majority against the Tutsi minority.[22]

In-Group Contradictions in U.S. Politics

One of the key patterns related to in-group bias is that we often create double standards for behavior and thinking. It's okay when *we* do it, but it's absolutely bad, wrong, unethical, or dangerous when *they* do it. This has been extremely apparent in politics, especially in the United States.

In-group/out-group dynamics are always part of any political system, but in the U.S. this dynamic escalated over the last decades and reached a fever pitch during the Donald Trump presidency of 2016–2020. But one episode that stayed with me occurred in the months leading up to the 2016 election, when Hillary Clinton was vying for the U.S. presidency against Trump.

A little social experiment conducted at the 2016 Democratic National Convention illustrated the in-group/out-group dynamic beautifully. Die-hard supporters of Hillary Clinton were questioned about how they felt about specific statements she had made during the campaign.

A middle-aged white woman wearing a blue "Madam President" T-shirt was asked, "What do you think about this quote from Hillary? 'I see improved relations with Russia from a position of strength only is possible.'"

The woman answered plainly with a shrug, "I agree with Hillary."

An elderly Black man wearing a cream-colored three-piece suit was presented with this direct quote from Clinton: "Immigration is a privilege, and we should not let anyone into our country who doesn't support our communities—all of our communities." He confidently responded with, "I virtually have to agree with that."

The next person in the video clip was a white man wearing a straw hat. He was told, "Hillary said, 'In order to achieve the American Dream, let people keep more money in their pockets and increase after-tax wages.'" Thinking on the spot he pieced together his response: "Yes. It's a great idea so we can start to move everyone up... and move people out of poverty."

In a quick follow-up, the interviewer inserted, "And this runs in severe contrast to what Donald Trump is proposing?"

"Absolutely," Straw Hat Guy retorted confidently. "Many of her comments run in severe contrast to Donald Trump."

But Straw Hat was wrong, as were the others.

Unbeknownst to these Clinton supporters, the quotes they had agreed with were actually statements made by Donald Trump, and this was a segment from the comedian Jimmy Kimmel's show.[23] As Kimmel said in the introduction to this informal experiment, "For the most part, I don't think it matters *what* a candidate says, it matters *which* candidate says it."

And he's right.

Other Democrats were shown to have agreed with Trump comments that were sexist, against public education strategies like Common Core, and even forgiving of an anti-gay marriage position, when they believed they were coming from Clinton.

In-group bias frequently manipulates us unconsciously to make exceptions for our own group. On top of that we find it very difficult to see these double standards or notice when our actions contradict our beliefs. This form of in-group-centric behavior has been demonstrated on both sides of the political spectrum. There were many contradictions demonstrated by Republicans in the U.S. during the Trump era; I'll highlight just a couple. Republicans have always prided themselves on "family values," emphasizing Christian morality, the sanctity of marriage, traditional gender roles, and heterosexuality, to name a few. A vivid example is that the Republican-controlled Congress impeached President Bill Clinton for having an extramarital affair while he was in office in 1998. Yet the Republicans were willing to ignore Trump's affair with Stormy Daniels, an adult porn star to whom he paid $130,000 in 2016 to stay silent about their relationship.[24] They dismissed as "locker room talk" the *Access Hollywood* tape released that same year in which Trump is heard bragging about being a celebrity and how it gives him the permission to sexually assault women—to kiss, grope, and even "grab them by the pussy" without consent.[25] And Republicans did not believe at least twenty-six women who accused Trump of multiple incidents of sexual misconduct, including allegations of rape, since the 1970s.[26]

Republicans were silent or aggressively defensive when evidence emerged of Russian interference in the 2016 U.S. election that was tied to Trump's campaign. Republicans have historically defined themselves as extremely suspicious of Russia, the antagonist of the U.S. during the Cold War era, which lasted from 1945 to 1991. "Communist" was and still is used as a dismissive slur for Democrats or a label for policy suggestions regarding the expansion of social welfare programs. Yet, in spite of this, they seemed to express little concern when the Russian government worked aggressively to sway the results of an election in favor of a Republican candidate.

The Republicans also pitch themselves as the party of "law and order," yet were willing to overlook the extreme level of corruption and criminality connected to Trump and his inner circle. Special

Counsel Robert Mueller—a bona fide Republican himself—indicted, convicted, or got guilty pleas from thirty-four people, including many senior members of Trump's team, on charges ranging from lying to Congress, lying to investigators, tampering with witnesses, obstruction of justice, hiding millions of dollars, and lying to banks to get loans.[27] These insiders included Trump campaign chair Paul Manafort, National Security Advisor Michael Flynn, campaign aide Rick Gates, campaign advisor George Papadopoulos, Trump's personal lawyer Michael Cohen, and longtime Trump confidant Roger Stone. Entire political careers have been ruined with a fraction of offenses connected to a leader, yet the vast majority of Republicans stayed firmly behind Trump. The willingness to ignore, deny, or defend such lawlessness is an extreme, head-spinning example of in-group bias.

In the end, the attitude of many Republicans in their unwavering support for Trump can be boiled down to a T-shirt slogan that went viral from a couple of Trump supporters: "I'd rather be Russian than a Democrat!"[28]

And this is why the level of polarization has become so dangerous in the U.S.: citizens of the same country have such anger, resentment, and distrust toward their fellow citizens that they would rather identify with an enemy state that is seeking to undermine their own country. There was a time when "country before party" was the norm, but not anymore.

And in the U.S., the political division is a racial one.

As a 2020 analysis of partisan leanings by Pew Research demonstrated, among 320,000 registered voters, 49 percent affiliate as Democrats, 44 percent as Republicans, and the remaining 7 percent as independents. White voters made up the majority of the Republican voters at 53 percent, compared to 42 percent of Democrats. Racial minorities made up 40 percent of registered Democratic voters, compared to only 17 percent of Republicans. Overall, white voters have been going down in numbers over time, making up 69 percent of voters in 2018–2019 compared to 85 percent of voters in 1996.[29]

Over time, the political divide has increased as U.S. liberals have become more liberal and conservatives more conservative. According to researchers, it's important to recognize that the division is not equal, as Republicans have become more extreme in their positions than their Democratic counterparts,[30] which is called *asymmetrical polarization*, a term that acknowledges that the divisiveness is not equally distributed. For example, after the election, most elected Republican lawmakers would not publicly acknowledge President Joe Biden's election win. Many continued to push the conspiracy theory that the Democrats had stolen the election from Trump, in spite of numerous courts, including the U.S. Supreme Court, throwing out their baseless, evidence-free allegations.[31] This led to the violent attack by a pro-Trump mob on the Capitol Building in Washington on January 6, 2021, which desecrated the building, terrorized the lawmakers inside, and left at least five people dead.

It's also worth noting that the last time such extreme Us/Them dynamics surfaced in a U.S. election was in 1860 when southern states refused to accept that Abraham Lincoln, the explicitly anti-slavery Republican candidate, had won the presidency. This was a key step toward the infamous American Civil War (1861–1865), still the bloodiest in the nation's history.

Although it's beyond the scope of this book, it's worth mentioning asymmetrical polarization, because solutions developed to bridge the political and racial divide with a "both sides are equal" approach distort our perception of the problem and power dynamics involved, skewing possible outcomes from the onset.

Inner Skill 4: Empathy

Categorizing is normal human behavior. So is a tendency toward anxiety, fear, and autopilot when dealing with new and challenging situations involving out-group members.[32] But unconscious bias and stereotypes are destructive, especially as they impact historically nondominant groups. So what can be done about the tendency to categorize? How do we reduce the Us/Them tendency that seems influenced by both genes and culture?

Empathy is a good starting point.

Empathy is the ability to tune in to the emotions of others, perceiving their feelings, needs, perspectives, and concerns. Although we can never truly "know" another person's experience, nor actually "walk a mile in their shoes," empathy will take us a distance down that road. Empathetic people are regarded as excellent in meeting the needs of others, in their personal or professional lives.

Emotions expert Daniel Goleman describes empathy as follows:

> a critical skill for both getting along with diverse workmates and doing business with people from other cultures . . . [where] dialogue can easily lead to miscues and misunderstandings. Empathy is an antidote that attunes people to subtleties in body language, or allows them to hear the emotional message beneath the words.[33]

Developing empathy for others is intimately linked to the inner skills of self-awareness and self-regulation. As the following example demonstrates, we can be ambushed by the Us/Them tendency in subtle and unexpected ways. But training to catch ourselves in the act may be helpful in expanding the circle of "we."

WE COULDN'T SEE EYE TO EYE

I was having a hard time focusing during a lunch-and-learn session on fundraising for the office charity. There was something odd about the mannerisms of the manager, Giselle, who was presenting. It was her eyes—they seemed almost closed the entire time she spoke to our group of thirty staff. Although Giselle held her head up and smiled in a friendly manner, her eyes seemed to be looking directly down at her feet.

It was very off-putting, at least for me.

Watching her, I began to feel uncomfortable and even a little agitated. Many thoughts rushed through my head: *Is she aware she's looking down so much? I wonder if others have noticed? Of course they have! How could they not? Surely she must know—someone must have given her feedback on this before they promoted her to manager status. Should I talk to her afterward? Is it my place to?* And so rushed the stream of mental blabber...

I became aware of my inner chatter and discomfort. So I took a breath and started sorting through my feelings. I became aware that I felt embarrassed *for* her, and that my feelings were strong enough that part of me wanted to leave the presentation. Her unusual mannerism was causing me to subtly tilt away from her. I also noticed the desire to turn to the person next to me and gossip about what we were witnessing.

Instead, I paused and noticed what was happening in my body. To myself, I quietly named the discomfort, as well as the desire to leave and to gossip. Doing this self-reflection somehow helped me shift gears. It allowed me to become curious about myself as well as Giselle.

I realized something: closing her eyes during a presentation was out-group behavior. She was breaking the social norm of what I considered a "good presentation." And as discussed, witnessing someone breaking a social norm is often a very emotional experience.

I further reflected on my previous interactions with this manager. During my consultancy in this large financial firm, I had talked to her a few times. She had avoided making eye contact during our conversations, too. I was, however, unaware of how severe it was until she was in front of the group.

Then it struck me: perhaps she was really nervous and extremely shy.

I hadn't considered this before, because my feelings of discomfort had overpowered my ability to think. This understanding helped me move past her out-group behavior and open the door further to empathy. *In spite of her shyness, she's giving a presentation in front of a group. She's pushing herself into things that are difficult for her? Okay—that's gutsy.* Feelings of admiration, understanding, and compassion now mixed in, muting the discomfort I originally felt.

I remember looking around the audience to see how others were reacting to Giselle. I did a quick scan of body language and found little that gave away how anyone else was feeling. This was normally a very pleasant staff team, cordial and friendly, and, true to form, they appeared to be listening attentively.

Perhaps it was just me? The presentation ended and we all went our separate ways.

In a conversation with a couple of staff members on a later day, Giselle's name happened to come up. An instant reaction popped like the top of an overshaken soda bottle: "She is so weird! Her eyes were closed the whole time!" said one, referring to the presentation. The other responded, "I thought maybe she was blind–I really did! How did she become a manager? Isn't speaking in front of others part of the job?"

Clearly I was not the only one who had noticed her out-group behavior. I was also quite certain that their charged comments specifically stemmed from Giselle's presentation. She was new and they didn't work with her directly. "I think she's painfully shy," was all I offered, in a nonjudgmental manner.

"Oh," responded the first person, whose face and tone softened. Apparently that hadn't been considered. The second person mumbled something and turned away. Although I can't be certain of the exact impact, their body language suggested that my words had altered something in them. I believe Giselle went from being someone "weird," whose behavior seemed difficult to understand or predict, to being "shy," a state her colleagues could better relate to and empathize with.

If the reaction of discomfort I shared with these two colleagues was an indicator, Giselle had been cast into an out-group based on "preferred body language during presentations." All three of us had in common a pile of feelings about Giselle that had led to an immediate negative judgment of her as a person. The important difference was that I had caught myself in the act of ungenerous thoughts and feelings, and so was able to shift something internally (and, perhaps, for the other two as well).

As we've discussed, if you can name it, you can tame it. By noticing what's happening to us, we engage our thinking mind. This can take some steam out of unconscious processes that push us toward behaving less generously. Curiosity can help foster understanding—both in ourselves and others—and empathy can lead to a deeper acceptance of differences, especially when interacting with out-group members.

TOOLS FOR CULTIVATING EMPATHY

Researchers C. Daniel Batson and Nadia Y. Ahmad from the University of Kansas have found that empathy can be developed through a variety of methods, including the following:[34]

- media such as books, TV, movies, or internet sources that offer perspectives of those different than ourselves
- intercultural dialogues between conflicting groups, as well as peace camps and personal storytelling opportunities
- discrimination-simulation activities that allow participants to experience a role that is marginalized
- specific educational programs designed to enhance positive feelings toward others

Cultivating empathy starts with deep listening skills and curiosity, both for us and for the person or group with whom we are interacting. I've found the following questions offer a starting point to enhance self-pattern awareness regarding empathy generally:

- How well am I listening to what the other person or group is saying?
- What are the feelings underneath the words being said?
- What emotions and judgments am I experiencing as I listen? (Name them.) How might these feelings and reactions be getting in the way of really understanding what's being said?
- How might it feel to be in their position, under their circumstances? What might they be needing or feeling right now?

More specifically regarding race and difference, the following questions may be useful to reflect on, especially when interacting with out-group members:

- What social identities—theirs and mine—might be at play in the interaction?
- Is there an interplay in this situation between dominant and nondominant identities? How might this help or hinder the communication?
- If the person or people are part of my out-group, how might that be limiting what is being heard or interpreted, by them or by me?
- Can I catch myself in the act of bias, assumptions, or stereotypes during this interaction?
- How might my lack of historical knowledge facing this group be getting in the way of understanding the person, group, or issue? Have I named this lack?
- What is their experience of me at the moment (how I'm listening and what I'm saying)?
- What respectful questions can I ask to better understand the person, group, issue, and circumstances?
- What might it be like to walk in this person's shoes for a day?

Finally, what new learning from this chapter helps expand your suite of racial pattern recognition templates? What may help grow your psychological pattern recognition tool kit?

Empathy functions in tandem with other inner skills such as self-awareness, mindfulness meditation, and self-regulation. Studies have shown that those who have a mindfulness practice can more accurately assess sensations within their own bodies—such as sensing their own heart rate—and are also more accurate at sensing what is happening for others emotionally.[35] And greater awareness of what happens in our bodies is a key part of the practice of mindful meditation. Further, enabling the power of curiosity requires self-regulation, especially if the interaction is emotional in nature.

Power: The Dividing Force

Marco's Rebellion

Marco began to hate the police. Again, they had pulled him out of his community and embarrassed him in front of his peers. The last couple of times, it was for minor infringements like stepping outside the town's boundary and mouthing off. I wondered what he would do this time.

Jail would be another round of humiliation. Last time he had broken out and made a mess of things, running through the three neighboring communities yelling about conspiracies and injustice. Many people didn't even look up. He was ignored.

Part of me felt bad for him. He had no idea of the extent of the prejudice or how significantly the odds were stacked against him.

Next, the police conducted a "search and seizure" in Marco's community, following up on rumors that drugs were being sold openly. Residents were made to stand like criminals, watching while law enforcement agents ransacked their meager buildings and homes. This after they had just rebuilt from flash floods.

Yet again, most people from the neighboring groups didn't even look up. Marco and his community, Orangevale, were once again ignored.

Marco and the Orangevale residents were getting seriously demoralized. I could see a sense of helplessness setting in. Marco would soon have others join him in his rebellion. Then things would really get chaotic.

The results are consistent. I know. I've done this many times.

This is the *City Game*, an experiential activity we use in our leadership programs to teach about the distribution of social power in society. We explore the concepts of *privilege*, unearned advantage or status based on one's identity, and its opposite, *marginalization*, disadvantage or low status based on identity. I've conducted it at least two dozen times in five different countries with young adults aged seventeen to twenty-four, and while participants always have fun and take part enthusiastically, the results are always the same.

It works like this. Fifty to seventy participants are divided into three groups and given color-coded team names like Orangevale, Redvale, and Bluevale. They are told the City Game is a competitive team-building activity. Their goal is to develop the best model city using nothing more than paper, tape, scissors, and their imagination. This, of course, is not the true learning objective of the game, which is not revealed.

With everyone in the same large room, masking tape marks physical boundaries, giving each team just enough room to stand in a circle. In this tight space, team members have to collaborate to design and develop their ideal model community. Winners and losers are announced at the end of the game.

To ensure protocol is followed, peer facilitators are given roles. There is the Mayor's Office, which has to approve all plans and budgets, and decide who the winner of the competition is. A number of people are in the role of police, to ensure that the rules are followed. The teams are given ninety minutes to complete the task. As the game director, I introduce the game and help it run smoothly.

What participants are not told is that the game is completely rigged. It is predetermined which community will do well and which will not. In this case, team Red—which also has fewer people than average—will receive the most support. Orange—which has the most people—will encounter the most obstacles, and Blue will fall somewhere in the middle in both size and support.

Red will receive approval on the first plan and budget they submit, regardless of what it is, and begin building their city within the first ten to fifteen minutes. The plan Orange submits will be sent back several times, so that they will be unable to begin building for about thirty minutes. Blue will be somewhat delayed in their plan, but not as much as Orange.

The police will start harassing the Orange team early, from asking them to make sure they aren't stepping on the tape boundary to taking one or two people to the designated jail. In jail the "rule-breakers" will be forced to undergo "punishments" like doing push-ups, singing songs, or reciting goofy poetic apologies for a few minutes. It's mostly in jest, but after a while, it starts to grate on the nerves of those who, like Marco, are regularly targeted. On the other end of the spectrum, the Red team gets accolades for whatever they do, with easy access to resources (extra tape, markers, and so on) and direct visits from the Mayor's Office. Team Blue experiences some setbacks to their plans, but not as many as the targeted group.

As game director, I act in a devious manner that maximizes the dynamics of privilege and marginalization. I am duplicitous in what I say and do, especially to the victimized team, Orange. I act concerned when I hear charges of "police brutality" and "unfairness," promising to do something to follow up with the Mayor's Office and police. But since the conspiracy is predetermined, there is no real improvement. To make a show of authenticity, every once in a while I will rebuke a police officer or get someone out of jail.

The heat gets turned up about halfway through the game, with the victimized Orange community experiencing everything from natural disasters to drug raids and rezoning, which reduces the size of their already overcrowded space. So even when this large group is just starting to get somewhere, some incident sets them back.

The result by the end of the time period is always the same. The privileged group, Red, has a very well-organized town that looks fabulous, usually filled with business districts and residential areas, hospitals, schools, community centers, airports, amusement parks, and even beaches. The middle group, Blue, has something decent looking, but not nearly as sophisticated as the Red town. And the group targeted for poor treatment, Orange, is usually sitting in a heap of crumpled papers, with some members running amok and screaming for revolution.

If a mentor or teacher happens to be in an observer role in the room, they usually find it difficult to not laugh out loud at the zany antics and unfolding chaos. Everything is so obvious, it's very difficult to maintain neutral facial expressions.

So what's the point? What can this exercise possibly teach us about privilege and power, when it's so grossly manipulated?

Although what's happening is obvious to observers, it isn't to participants. Most of the participants don't realize what's going on. And that's the essential lesson.

Recall, they were told this was a competition. So they are focused on the "purpose" of the game: to build a model city (distracted by the specific moment of their lives, as we all are). It's after the City Game is over, during the debriefing session, that the true learning happens.

All of us coconspirators remain in role, even when the bell indicating the end of the game sounds. With each group seated around their model town, the Mayor's Office proclaims who came in first, second, and third, with applause and cheering.

To spice things up for the debrief, one team at a time is invited to walk around to look at the other groups' efforts. This amps up the emotional wattage in the room. The first-place team feels pride and success, showing off their fabulous city. In contrast, the third-place team, with their tattered shantytown, experiences momentary embarrassment, shame, and even anger.

After the groups have seen each other's cities, I ask each team a question: "What were the factors in your success or failure?" It's the contrasting reflections between the top and bottom teams that are most interesting.

The first-place team almost always says the same things. Their success was due to creating a plan, dividing up the tasks among their members, and working collaboratively to build their winning creation. They are genuinely proud of their accomplishments.

And, as a group, they almost never have a sense that they were given any advantages by the Mayor's Office or police. They generally believe they got the top prize because of their hard work and merit. Remarkably, they rarely notice anything else happening in the room, in spite of the chaos impacting the persecuted team.

The members of the last-place team, Orange, usually offer a different perspective. Many tend to feel that they were unfairly targeted and regularly harassed by the police for small infractions. They identify the drug busts and natural disaster that affected their community, wiping out most of what they had managed to build. It's also common for members of the group to say that they, themselves, were to blame—that they didn't work well together, they had poor leadership, or they had some disruptive team members.

And then the reveal happens.

We drop our masks and tell them the game was fixed. Again, the results are predictable. The group targeted for poor treatment is jubilant—*we knew it!*—while the privileged group is crestfallen and often silent. The ensuing conversation is rich. Together we reflect on how this game relates to real life, making the links to both privilege and marginalization.

Having run this activity so many times, I've learned some important transferable insights. I'll highlight three. First, regardless of who we are—our identities—it's hard to see privilege. Even during an experiential activity as overtly manipulated as the City Game, those who receive preferential positive treatment most often don't detect it. The reality is the same for all of us because we don't feel our privilege when we have it, even though it may seem obvious from an external perspective.

Second, when we are in the privileged group, we are focused on our own hard work and challenges. So it's difficult to see how the system functions to reward our efforts while holding back those of others based on such flimsy factors as social identity. (Recall the resumé studies demonstrating pro-white-name bias. How many people with white-sounding names would be aware of this significant advantage in the workplace?)

Finally, during the City Game, we are always conscious to create integrated groups. Whenever possible we mix people who, in real life, come from privileged and marginalized backgrounds based on factors such as race, gender, or class. Nonetheless, the results are consistent; the actual lived experience of the participants factors little in the process. If we are placed in the group targeted for oppression, it's common to feel negative, reactive, and helpless.

If treated as privileged, we are engaged and successful because we are invested in "the game" and its outcomes. Even participants who are disruptive or oppositional during the leadership program often settle down and focus when they are placed in the privileged group.

So what factors are at play here? How can we understand this last dimension of Deep Diversity—power? How do social power patterns create dominant and nondominant groups in society, groups that, as a result of privilege or marginalization, can face very different realities and have strikingly different needs? The previous three components of this framework—emotions, bias, and identity—emphasize the similarities of the human condition, how we are the same. Not so for power. Power—the distribution of socioeconomic, political and cultural power—is what entrenches differences between groups. Power determines who is centered and who is not, who gets defined as "Us" or "Them." And this kind of all-pervasive power is the key ingredient that escalates individual acts of discrimination into systemic problems such as racism or sexism. The framework provided by social dominance theory will help illuminate how the patterns of privilege and power in the City Game are magnified and replicated across society.

Systemic Discrimination

Psychologists Jim Sidanius of Harvard University and Felicia Pratto of the University of Connecticut conducted studies and synthesized research from countries around the globe. They uncovered something very important about human cultures and, perhaps, our intrinsic nature: that social hierarchies exist in all nations, creating inequality and serving as the source of most forms of intergroup conflict.[1]

They drew on research from countries including the United States, Canada, Mexico, the United Kingdom, Germany, the Netherlands, Sweden, Israel, Palestine, Kenya, South Africa, Russia, India, China, Taiwan, Japan, Australia, and New Zealand. Regardless of where we live, privilege and marginalization seem to be an integral part of the human experience.

Sidanius and Pratto's framework, called *social dominance theory*, states that dominant groups exist in each society. A group receives privilege based on a relatively arbitrary value such as race, ethnicity, clan, caste, gender, or sexual orientation due to historical, economic, and political reasons specific to that context. There is no innate or natural reason for such dominance to occur—no one social group is more capable or deserving than another. But it happens. It's always happened. It seems to be something in our human nature.

Usually a group at the top of the power pyramid is granted automatic privilege—unearned higher status based simply on their group identity. All other groups—usually defined by a "minority" status—exist in a hierarchical order below and experience a greater or lesser degree of marginalization. Sidanius and Pratto's research shows that social dominance is based on variables that are specific to each context and consistent over time. They argue, for example, that in Europe—a place known for castles, kings, and commoners—class is the historically dominant force.[2] Although race plays a significant part, they suggest it's a lesser one. In contrast, in the United States and Canada, race is the most consistent factor. Class plays an important but secondary role. The two researchers have identified a variety of ways in which hierarchies are supported in society, including the following:

- individual acts of discrimination
- institutional discrimination
- psychological distortions
- self-destructive behaviors
- cultural myths

These five factors can serve as a basic frame to help expand our racial pattern recognition framework by illuminating the issues of power, privilege, and marginalization.

Individual Discrimination

The Latin roots of *prejudice* mean "to judge in advance." It is often described as a preconceived idea or belief about individuals and groups that may or may not be based on reason or reality.[3] If prejudice is the belief, discrimination is the related action. And it is the small and large daily acts of discrimination, both conscious and unconscious, building up over days, years, and generations, that support and maintain a power imbalance between dominant and nondominant groups.[4]

We can think of many individual or private acts of discrimination. A landlord won't rent out an apartment because of the applicant's race. A police officer stops more Black drivers than white. The manager of a grocery store hires only women as cashiers. Educators and employers have lower expectations of people with disabilities. And most of us tend to make less eye contact with out-group members.

Institutional Discrimination

A number of years ago, I worked on an intercultural dialogue project in the Netherlands. I was the houseguest of a judge there named Hanneke. We discussed at length how unconscious racial bias might impact the Dutch criminal justice system. It might help explain the overrepresentation in the system of Moroccan, Turkish, and Caribbean youth, groups that were at the bottom of the social hierarchy. This was a relatively new concept for Hanneke, and our conversations weren't easy. During one discussion, however, she shared an insight that had deepened for her since my last visit to the country:

> When I look into the eyes of a Moroccan youth whose case I'm hearing, I don't really see anything. But this is different than when I look into the eyes of a [white] Dutch child standing in my court, as I see

something behind it, something familiar. I feel something that I don't with the Moroccan child.

I was struck by the honesty and simplicity of her insight. It cut to the heart of the issue. *It's easier to have compassion and understanding for those who are most like ourselves.* Hanneke had expanded her racial pattern recognition skills by becoming aware that her decisions as a judge were not as "objective" as she once believed. Fairness was threatened because she experienced more *feelings* for those from her racial in-group. Until our conversations, she had never really considered patterns of implicit bias. Nor had she been motivated to catch herself in the act, so it had remained in her unconscious. If a majority of a country's judges are white and lack bias racial pattern recognition skills, what might the impact be across the entire criminal justice system?

This is one critical way discrimination is institutionalized and becomes systemic. Leaders hold positional power and make choices that are more favorable toward members of their own group—in-group bias—usually without awareness that their choices might be prejudiced. And extra help or concern for our in-group members is what researchers Mahzarin Banaji and Anthony Greenwald suggest is a greater problem today than hurting people from out-groups.[5]

For example, the majority of judges in Canada have historically been, and continue to be, white. During appointments made between 2016 and 2020, only 13 percent of judges appointed by the Canadian federal government identified as Indigenous or racial minorities, up from 2 percent of appointments from 2008 to 2012.[6,7] This situation was exacerbated by a selection process that had no transparency or clear criteria, infuriating groups that represented lawyers who identified as Black, Indigenous, or people of color. They felt their members were regularly passed over for such appointments, in spite of outstanding qualifications.[8]

Institutional discrimination can be identified by whether, over time, organizational decisions result in a disproportionate allocation of social, political, and economic wealth and benefit to certain groups over others, whether that occurs consciously or unconsciously.[9] We have to ask, are all social groups treated fairly by institutional decisions and processes? The answer is an unequivocal no.

The appointment of judges is one small piece of evidence in long-standing patterns from within the criminal justice system. Mountains of data from decades of research demonstrate that dominant groups benefit more from many policies and practices than nondominant groups do. The Sidanius and Pratto research shows that this is a global phenomenon that is evident in any nation in which it has been seriously studied.[10]

The research unequivocally shows that nondominant group members—whether they are Black people in the United States, of Moroccan ancestry in the Netherlands, of Korean roots in Japan, of Arab background in Israel, or the Aboriginal people of Australia—are discriminated against in all stages of the criminal justice process. This applies from the likelihood of arrest and severity of charges filed, right through to higher rates of convictions and harsher sentences and punishment.

As with all aspects of an oppression such as racism, there is a deep history of mistreatment within criminal justice. For example, the U.S. continues to hold the global record for imprisoning more people—both per capita and in absolute numbers—than any other nation including autocratic states like Russia and China.[11] And the U.S. prison population is highly racialized. Blacks and Hispanics make up 33 and 23 percent of prisoners—56 percent total—compared to making up only 12 and 16 percent, respectively, of the overall adult population. White adults make up only 30 percent of the prison population but 63 percent of the general population.[12] Such systematic racial discrimination did not occur overnight; it is tied to generational patterns that stretch back to slavery, when Blacks were relegated to subhuman status and legally considered only three-fifths of a person. According to Jennifer E. Cobbina, professor in the School of Criminal Justice at Michigan State University, "slave patrols were among the first state-sponsored police forces and controlled the slave populations by restricting Blacks to certain places and monitoring their behaviors." After slavery was abolished, new discriminatory laws were erected in the late 1800s called "Black Codes" which "created new offenses, such as loitering and vagrancy, to be punishable by fines, imprisonment and forced labor . . . applied selectively to Blacks by all-White police and state militia forces."[13] This and other legislation under the umbrella of Jim Crow laws established different rules for whites and nonwhites, reinforcing deep racial segregation such as white-only neighborhoods, restaurants, and water fountains, and back-of-the bus status for people of color. Those who tried to defy such laws were faced with various forms of punishment including fines, arrest, jail time, violence, and sometimes death.

And those historical legacies show up as invisible influence today, ghosts that still haunt the criminal justice system, influencing who deserves punishment or leniency. The pattern plays out repeatedly. A stark example from a few years ago is that of twelve-year-old Tamir Rice, a little Black boy seen in security camera footage playing with a toy gun in a small park in Cleveland, Ohio. Alone and seemingly bored, he kicked at some snow stuck to the curb until a police car zoomed into the scene and an officer jumped out and immediately shot the child, believing Tamir to be holding a real gun. The video showed there was no pause, questioning, or consideration.[14]

In contrast, a month later that same year, a forty-three-year-old white woman, Julia Shields, drove around an upscale neighborhood in Chattanooga, Tennessee, dressed in body armor, using a real gun to fire shots at random people and cars, as well as pointing her gun at a police officer and children. Police gave chase by car and on foot, and when she was apprehended, it was without incident or injury.[15]

Sidanius and Pratto conclude that criminal justice systems around the world are heavily skewed, both consciously and unconsciously, to protect the interests of dominant group members. It may be difficult to accept that the criminal justice system serves as a de facto mechanism to keep groups "in their place" and maintain the status quo. But when we view society through the racial patterns that emerge through the experiences of millions of people, both current and historical, that conclusion is difficult to deny.

NO NEED FOR A CONSPIRACY

In 2021, the U.K. government released a report prompted by the previous year's George Floyd protests, which concluded there was *no* evidence of institutional racism in the country and suggested that while there were some isolated problems, Britain was a post-racial society because so much had improved over the generations.[16] When looked at more closely, the authors behind this controversial commission—self-described middle-aged, "successful minority professionals"—displayed an old-fashioned understanding of racism, which holds that for racial discrimination to be systemic there must be evidence of *intention* to exclude ethnic minorities.[17]

In the context of egalitarian countries like Canada, the United States, and the United Kingdom, institutional discrimination exists without intent or undercurrents of conspiracy. Unlike in the past, no shadowy cabal of overt racists attempts to keep Black, Indigenous, or people of color communities down—there doesn't need to be one. The unconscious drive to favor those most like us, our in-group bias, is enough. All that's required is for a majority of leaders to be from the same in-group. Unconscious bias patterns that privilege the dominant group will flow through the system, like water running downhill.

We can do this without realizing it and in spite of our explicitly stated values about equality and democracy. The social hierarchy is maintained through individual leaders making decisions, enacting or following rules and procedures across corporate, public, and not-for-profit institutions, including the courts, police, media, health care services, schools, banks, big business, sports, arts, and governmental as well as religious organizations. And, as in the case of the 2021 U.K. racism review, using like-minded racial minorities to validate the leaders' version of progress and worldviews regarding systemic forms of discrimination.

Of course this does not mean that, due to their identity, 100 percent of the people are affected by privilege (or marginalization) 100 percent of the time. But patterns of unfair treatment based on who we are happen with enough consistency, creating significant social disparities between groups. This fosters distrust among nondominant group members and holds back the lives of many hardworking individuals and communities. Further, such bias and favoritism function in today's society to reinforce hard-to-see cultural and social patterns that are powered by the underlying momentum of history.

Psychological Distortions

Part of my personal story involves desiring to be white when I was growing up, to the point that I avoided kids of my own South Asian ethnicity. Called *internalized racism*—the incorrect belief that our group (its norms, culture, and people) is inferior—this is one of the potential psychological impacts of power on nondominant group members.

In the City Game, some participants in the group targeted for unfair treatment spoke about this feeling—that their team did not have enough leadership or teamwork, that their situation was somehow their fault. The seeds for self-doubt and insecurity were taking root, even within a simple ninety-minute activity.

We've known about this effect for a long time.

A famous 1947 U.S. study demonstrated that a majority of Black children preferred to play with white rather than Black dolls. Researchers found that the Black children felt that the white doll was not only more attractive, with nicer skin color, but also "good." The Black doll, in contrast, was considered ugly and "bad." Interestingly, the white doll was preferred even more by Black kids in integrated schools in the north than by their peers in segregated schools in the south.[18]

Although that study took place more than a half century ago, when I described it to education leaders during a conference more recently, a diversity officer from a school district in southwest Ontario shared how accurate those results still were. As a standard practice, the administration provided dolls of different races to kindergarten classes. Over the years, they consistently found that white dolls needed to be replaced regularly—indicating high use—whereas dolls of color were preferred less for play by kids of all backgrounds, including in schools with large populations of students who identify as Black, Indigenous, or people of color.

Although many may argue that all groups are equally *ethnocentric*—that we prefer those from our own in-group—that is only partly true. Power, our group's status, is a major factor in how we view and value our own group compared to other groups.

Internalized dominance—the incorrect belief, conscious or unconscious, by dominant group members that their group's values, norms, culture, and people are superior—is the story from the other side of the coin. For whites, the dominant racial group in European-based nations such as Canada, the U.S., or the U.K., a large body of research demonstrates greater levels of ethnocentrism and in-group bias than among those who are nondominant. For example, in North America, white people tend to show a greater average pro-white bias than Blacks show pro-Black bias or those of Asian ethnicity show a pro-Asian bias.[19]

Although many studies indicate that pro-Black bias, for example, has increased within the Black community over the decades, that has mostly occurred at the *explicit* level (what people say about themselves).[20] Although this is important, Implicit Association Tests indicate that at least 50 percent of Black Americans still have an unconscious anti-Black, pro-white bias.[21] Like air temperature compared to ground temperature, explicit bias changes more easily than implicit bias, which takes a much longer time. And as discussed previously, our implicit bias is more often a more accurate marker of our collective behavior and choices than what we may explicitly say about ourselves.

In fact, in many cases around the world, there is a tendency for low-status group members to think less of themselves than of high-status group members. This has been found among the Maori of New Zealand, Ethiopian Jews in Israel, and Black children in the Caribbean, to name just a few examples.[22]

Self-Destructive Behaviors

It's also important to examine how we as individuals can contribute to our own marginalization through self-sabotage—the use of unhelpful thoughts and actions to cope with difficult feelings and existing problems.[23] I offer this next part of Sidanius and Pratto's framework with care and some trepidation. It's a part of the picture, but one that can easily be turned into blaming the victim. With this caution in mind, we can probably all recall choices that we or others have made that hindered rather than helped the situation. Here's one from the context of education.

I was a fairly new teacher when the words of thirteen-year-old Faith stunned me: "Like my mom, I'll probably be on welfare." This mixed-race child suddenly landed in my classroom one day in the spring, with a hard-edged attitude and unpredictable behavior. Some mornings, her face was open and clear and she would really engage in learning like any other student. Other days, Faith would turn up mid-morning, pale-faced like she'd been up all night, and bark at anyone who crossed her path, including me. Sometimes she didn't show up at all. During health class, her questions and answers suggested she knew a bit too much about sex for someone her age. When I collected homework, she would make an excuse to go to the bathroom or get into a fight with one of her classmates. Over time, I discovered that she would rather do this than feel incompetent or embarrassed in front of her peers. It was early in my teaching career, and I was learning from Faith about how an individual can hinder rather than help their own advancement. I was never certain whether her behaviors were the result of poverty, racism, violence at home, or something else. As an equity-oriented educator I was sensitive to the possibility of underlying patterns so I did the best I could to support Faith and accommodate her needs. Yet the classroom structure and my formal teacher training didn't offer many tools to help.

Two things were clear to me. First, that her self-harming behaviors outweighed her self-helping strategies.

And second, that Faith's negative words, choices, or actions were not innate but rather circumstantial. When we feel powerless or helpless, or when essential needs in our life are not met, we lash outward or inward—and often both.

This does not imply that life outcomes are predestined.

Even if our home or community context is the intersection point for poverty, racism, and violence, where the opportunities to fall into despair, negativity, and self-destruction are plentiful, individuals in such situations are able to get past self-destructive behaviors when they have key supports in place.

People like Javier Espinoza. Now a public speaker, he describes learning from his courageous mother, who took her four kids and fled her abusive husband and worked hard to make ends meet, taking menial, exploitive jobs.[24] He also credits the shelter his family stayed in and a mentor who helped him question where he was going in his life and how he might help others who were in similar situations.

Without such supports, self-sabotaging behaviors reinforce the institutional biases that keep many nondominant group members on society's margins from achieving their potential. Javier, against statistical probability, attended university. Today, he gives back to his community by working in the field of domestic violence.

As Javier's story demonstrates, given the right circumstances, we can make individual decisions to help ourselves. This is possible even in situations where cultural or societal pressure pushes us to make self-destructive, marginalizing decisions.

In the context of education, for example, Sidanius and Pratto show that some marginalized youth develop oppositional attitudes toward teachers and the education system in general. Like Marco experienced in the City Game, negative encounters with authority figures can cause people to behave in disruptive ways and lose interest in playing by "the rules of the game." Sidanius and Pratto point to studies focused on youth of color from inner-city contexts, which indicated that achieving academic success was sometimes stigmatized as "acting white." Teachers and school administrators were distrusted, due to perceived and real discriminatory practices. These marginalized youth considered rejecting academic success to be an act of resistance or defiance to a system of oppressive education.[25]

Laurence Steinberg's longitudinal study tracked more than 20,000 U.S. high school students from the mid-1980s, following them over a ten-year period. This research found that inferior academic performance of Latinxs and Blacks could be attributed to their cutting class more often, doing less homework, being less focused on their school, and being less engaged in academic achievement.[26] This dynamic between students from dominant and nondominant groups is repeated across the globe, including in Australia, Bangladesh, Belgium, Canada, the Czech Republic, Denmark, England, France, Germany, Kenya, Pakistan, Sri Lanka, and Sweden.

Sidanius and Pratto also suggest that academic success is influenced by whether youth believed educational achievement was linked to factors within their own control, such as hard work and effort.[27] In contrast, the more students felt success was determined by uncontrolled influences, such as native intelligence or early school experiences, the less well they did. Most curious about this finding was that the longer an immigrant group was in the United States, the more they began to believe factors were outside their control and the less well they did academically. Although this finding was more true for some groups (Blacks, Latinxs) than others (Asians, whites), it suggests something important. The longer we experience marginalization, the more psychologically vulnerable we become to negative beliefs about what we are capable of.

Beyond education, evidence suggests that self-debilitating behaviors are broader among nondominant group members than among their dominant counterparts.[28] Such behaviors include higher drug abuse, higher rates of child and spousal abuse, and a greater tendency to use extreme violence to resolve disputes. The research suggests these behaviors come from feelings of inferiority, self-loathing, and inner-directed aggression.

Recall, from Chapter 2, the overlap between social and physical pain in the brain, and that even a few minutes of exclusion during an insignificant cyber pass-the-ball game can result in feelings of low self-esteem and lack of control. Consider what might happen if we consistently experience being left out and stigmatized, the effect magnified over years, decades, or generations.

Self-destructive behaviors are thus directly linked and nurtured by institutional discrimination. In one of my trainings with high achievers in the media industry, Abdi, a child of new immigrants to Canada, shared that in spite of doing well at high school, he was encouraged by his guidance counselor to consider community college rather than pursue his dream of a university degree in journalism.[29] It is well established in the research that racial minority students—especially those who are Black and Indigenous—are streamed into courses below their abilities while also experiencing suspension and expulsion rates much higher than those of their white counterparts in Canada, the U.S., and the U.K.[30]

If it looks like members of my group are consistently being mistreated in and by the "system"—that we're on the wrong side of statistical probability—helplessness and hopelessness can set in. These feelings can give many self-destructive behaviors logical merit. What would possibly make me think I will be the one to defy the odds?

Ongoing patterns of low expectations, unfair discipline, and misrepresentation of culture are just a few of the ways that institutional racial discrimination can contribute to self-destructive behaviors. The result is Faiths who expect a future on welfare, Abdis with certificates from community colleges rather than university degrees, and inner-city youth of color cutting classes and dropping out without high school diplomas because they believe educational success is outside their control. In other words, negative psychological impact on nondominant groups is one more factor bolstering the hierarchical power position of dominant groups.

Consensual Cultural Myths

Deborah Barndt was the university professor who taught me Antonio Gramsci's concept of *hegemony*. She explained that hegemony is power wielded through ideology. Standing at the front of the class, she wiggled the fingers of one hand, pointing down like a puppeteer manipulating dolls on a string. "It's coercion from above," she said, "combined with consent from below." The fingers of her other hand pointed upward, wriggling like little dancing worms in response to the invisible strings from her top hand.

Sidanius and Pratto, summarizing the work of Gramsci and others, describe societies having enduring cultural myths that are believed by large numbers of people from both dominant and nondominant groups.[31] The consensual nature of cultural myths is another factor that reinforces hierarchical relationships between groups.

For example, in the United States, one of the oldest myths is the so-called American Dream—that socioeconomic and political success is available to anyone who works hard. The flip side is equally enduring: if you don't achieve social or economic success, you didn't work hard enough. This belief in *meritocracy*—that anyone can make it if they put in the work—has deep Protestant roots that lead back to the founding of the nation. This ethic is also strong in Canada as well as in the United Kingdom, where it originated.[32]

An extension to the American Dream myth is the belief that racial minorities have just as good a chance as whites do to get jobs, education, and housing. Gallup polls conducted in 1997, 2007, 2009, and 2019 confirmed that a majority of the U.S. population has a shared belief in this myth.[33] In the 2019 survey, 73 percent of whites and 62 percent of "non-whites" believed that it was possible that the American Dream was attainable "if you work hard and play by the rules."[34] It's safe to assume that if the category of "non-white" was broken down further, the most marginalized groups like Black or Indigenous people might skew below the average. Regardless of such differences, Sidanius and Pratto point out that what is most remarkable is that there is such *persistent agreement*, with at least half or more of the entire population believing in equal opportunity, in spite of generations of evidence to the contrary. That's the enduring power of a cultural myth.

And the American Dream appears to have been realized by many people, including, spectacularly, President Barack Obama. He overcame humble roots, including both racial and economic barriers, to win the most powerful elected position on the planet. His story and those of other successful people like Oprah Winfrey or Maya Angelou seem to have important psychological impacts on the rest of us.

For one, we tend to remember and be affected by stories of individuals rather than the plight of faceless millions.[35] Second, Obama and Winfrey, as positive exemplars, help justify—both consciously and unconsciously—a sense that the system is inherently fair because "they made it." The implicit suggestion is that minoritized people are successful (or not) entirely because of their own strengths or weaknesses. This perspective takes attention away from systemic patterns related to racism or poverty, for example. It's easier to remember the individuals who defy statistical odds, and as a result we can be lulled into complacency and not recognize the inherent structural dynamics at play.

Obama and His Preacher

In 2008, two race-related events caught my attention. Although they were separated by significant geography and context, they were linked in content.

Close to home, the public school board in my city announced that it would create an alternative school with a Black-focused curriculum to address disproportionately high rates of failure among Black students. This controversial decision had been over a decade in the making. A multilevel government task force on education had first introduced the idea. Whenever this issue was raised, heated debates about race followed.

Reacting to the decision, proponents of the school were excited and relieved that some positive steps were being taken to address a huge historical problem. If it was successful, perhaps learnings from this experiment could be transferred to regular schools? Critics of this initiative, some of whom were Black, were upset. They argued that "segregation" and "reverse racism" were not going to fix the problems.

Across the border, a higher-profile event was taking place. Barack Obama, prior to his historic win to become the first Black president of the United States, was still competing with Hillary Clinton for the Democratic candidacy. At the same time, the church to which he belonged—Trinity United Church of Christ in Chicago—came under scrutiny due to inflammatory comments made by its longtime pastor, the Reverend Jeremiah Wright.[36]

I was flipping channels one night and stumbled across a conservative TV personality, Sean Hannity, interviewing Reverend Wright about allegations that he was running a Black separatist church.[37] In the segment, Hannity quoted directly from the mission statement on the church's website, which explicitly stated:

- commitment to the Black community
- commitment to the Black family
- adherence to the Black work ethic
- strengthening and supporting Black institutions
- pledging support to Black leadership who embrace the Black value system

Hannity asked, "What if a church with a white congregation specified a similar racial focus in its mission statement—but replaced 'Black' with 'white'? Wouldn't that church be considered racist?"

Reverend Wright gave a complex (and somewhat confusing) response about liberation theology. Hannity pushed back and asked him to address the specific question about racism. The discussion devolved from that point, with both men talking over each other, at once defensive and aggressive.

Oh, brother, I thought to myself, *is this what passes for an interview these days?*

The TV clip, however, stayed with me. I thought Hannity had asked an important question (although I'm cynical of his motivations and sincerity; his modus operandi tends toward spectacle and polemic).

Hannity's core message was challenging: "I think as a Christian... you should not separate by race in this day and age." Another interviewee made the point that Trinity United's mission statement did not advocate Black superiority but rather restated the principles of self-reliance and self-help—the very values conservatives like Hannity profess. But I wasn't satisfied that the original question was fully answered. It's a question that comes up frequently in racially charged contexts.

The undercurrent to Hannity's argument was that we live in a post-racial, color-blind society, and that exhibiting Black pride—even under the banner of self-reliance—should be regarded as racially dangerous, like demonstrations of white pride by members of the Ku Klux Klan. Hannity's comments, the most powerful of the interview, were a hallmark of astute conservative political strategy: simple questions (and solutions) for complex problems.

Similar conversations occur about programs that seek to improve outcomes for historically disadvantaged groups, such as affirmative action (known as *employment equity* in Canada and *positive discrimination* in Europe). My impression is that a handful of people believe these programs are important and helpful initiatives, while another handful find them distasteful and discriminatory, describing them as "reverse racism." The rest fall somewhere in between these positions, with most remaining unclear about the need for such programs.

The central question can be distilled to this: In egalitarian, democratic societies, should minority groups be treated differently and given greater license on certain issues? Or should "they" always be held to the same standards as "everyone else"?

The Unanswered Question: White Supremacy

Looking at human groups through the lens of social dominance theory, we see that dominant and nondominant groups can have experiences of living in the same society that are not only very different but incredibly unequal. The buildup of individual and institutionalized discrimination over time combines with the belief in consensual myths, giving the dominant in-groups distinct psychological and material realities based on arbitrary social identities.

Although it is hard for many of us to accept in Canada, the U.S., and other European-based contexts, the net consequence is a racial hierarchy that was designed by, and for, white people. That's why some prefer to use the term "white supremacy" not just for overt bigots and hate crimers but to describe the overlapping social, political, economic, and cultural systems that perpetuate racism by one group—white people—who dominate decision-making and material resources.[38] Present-day patterns of racial disparities and supremacy thinking that privilege the white in-group more than people-of-color out-groups are tied to legacies of segregation, colonization, and slavery. (Again, this doesn't mean 100 percent of the people are affected by their racial identity 100 percent of the time, but entrenched patterns repeat with enough consistency with devastating social cost.) Because groups do not start on an even playing field, what they need to achieve their human potential is very different.

So, should Black-focused churches and schools be allowed when "white only" is not? I believe the answer is "mostly yes," with one key caveat, which I will return to.

Sometimes such identity-specific organizations are very needed. When a local institution like a church or school is created to serve a nondominant group, and explicitly states, for example, a commitment to the "Black family" or to "Black businesses," it's an attempt to help blunt the material and psychological impacts of white supremacy. Social spaces in which role models, positive stories, and institutional decisions are made with the nondominant group explicitly in mind can improve self-esteem and normalize our identities. This is particularly the case when the nondominant group is involved in the planning and decision-making. From the Deep Diversity perspective, such institutions open a space to create new habits—new neural pathways—that can reduce internalized psychological distortions about our group.

For a white person—the norm in North American/European societies—such seemingly small things are often taken for granted. They are part of the package of privilege. It is relatively easy to be immersed in environments where the white in-group is the bulk of the population and of the decision makers. It's usually a safe assumption that a white individual's school will have people who look like them, including peers, teachers, administrators, principals, and trustees. As a result, they may be unaware that the curriculum is built on white, European values and history. White dominance results in easy access to a multitude of role models and positive stories about white people. This is how privilege operates in daily life. Invisibly, white supremacist thinking becomes the default habit as neural pathways reinforce the belief that the dominant group members are "normal" while all others are cultural outsiders.

Combined with an invisible, unconscious racialized momentum within society, white people as a group progress more easily through schools, at work, and in community contexts and are better able to develop their human potential. That is less the case for many Black, Indigenous, or people of color who have to swim against the current or assimilate in order to advance, hiding or losing key parts of who they are in the process.

The crux of the matter is that creating a space for nondominant racial groups must be a conscious act. This explicitness is what gets some people emotionally worked up. It may look to some like a double standard because many people don't recognize this need to temporarily segregate as a necessary survival tactic. For the dominant racial group, however, society already is designed in their favor—racial privilege, unconscious and therefore invisible, is taken for granted. They can congregate in predominantly white-dominated spaces whenever they want.

In this historical period, it is significantly different for a church, school, or organization in Canada, the United States, or the United Kingdom to have a "Black-focused" mandate than a "white-focused" or "white only" mandate. In the current reality, white people are the majority population and hold the bulk of decision-making positions in society. Without consciously seeking it, if I am white, I can live, work, learn, and play in many, many places across all three nations, where decisions unconsciously and consciously are made to support me and my identity.

And when this reality is coupled with the trauma of North America's violent pro-white past, there is good reason to be wary of organizations explicitly wanting to be "white only." (And pro-white groups are quite different from cultural groups that celebrate their British, Irish, or Dutch ethnicities.) As the population balance changes, consciously managing shifting power structures becomes even more important. In Canada, the United States, the United Kingdom, and Europe, it is generally accepted that sometime in the second half of this century, the scales will tip and whites will become a minority. If we do not learn to share power, what will our lives look like with a white minority as the dominant group? Or if we don't learn from our experiences, will the tables simply be turned, with whites becoming marginalized?

When Segregated Spaces Are Okay

If girls-only classes result in better performance in math, then why not offer the option of sex-segregated classrooms? If regional health services with a specific Indigenous focus are more effective than mainstream medical facilities, why not offer more of them?

We've always known that one size does not fit all. That's especially the case for issues of race. Of course, I'm not talking about institutionally or state-enforced segregation, but about places where the underlying assumption is that we have a *choice* to join or not join such an intentional program or organization. There is an important positive side to optional segregated spaces.

Such environments, however, can also result in the promotion of hate rather than healing. That is the caveat to my "mostly yes" response. If groups work to promote violence, fear, and animosity—regardless of whether the groups are socially dominant or non-dominant—then I believe they are unhealthy and unhelpful, and should be stigmatized. In extreme cases, some organizations should be designated as hate or terror groups whether they are under the ever-expanding white nationalist umbrella like the Ku Klux Klan, Proud Boys, Atomwaffen Division, The Base, Patriot Front, or the American Freedom Party or one of the more rare clusters like the New Black Panther Party that attracts people of color through promotion of violence against white people and Jewish communities.

The issue then becomes who decides what is and isn't considered hate and violence. Again, it comes down to power. Who gets to define the problem and create the solutions? For example, if nondominant groups are critical, outspoken, or angry about issues like racism, sexism, or poverty, they are more likely to be regarded as a threat than others. Dominant group members, doing the same thing, usually have more leeway as a result of their privilege. We witnessed some very extreme examples of this in the context of the U.S. in 2020–2021. The graphic murder of George Floyd, a Black man, by a white police officer, went viral across all media platforms at the end of May 2020, kicking off widespread protests against anti-Black racism and police violence, not just across the U.S. but globally.

Shortly after, on June 1, peaceful protestors in support of Black Lives Matter were aggressively cleared from Lafayette Square in Washington by police officers clad in hard riot gear using rubber bullets, pepper spray, mounted horses, shields, and batons. Over three hundred people were arrested in a single night. There was collaboration between a variety of government of agencies in anticipation of the protests including the D.C. police, National Guard, and the Secret Service as well as the Bureau of Prisons, the U.S. Marshals Service, and Immigration and Customs Enforcement. Donald Trump even used the military to clear a path through the protestors so he could dramatically walk across the street to a church to hold up a Bible for a highly staged photo op in the midst of the protests.

This was a "massive over-response" by the police and state, according to experts in the field like criminologist Ed Maguire from Arizona State University.[39] And this scene was repeated across the nation in the following months at many post–George Floyd protests. For example, in Graham, North Carolina, a peaceful rally led by the local Black church that included children and the elderly was on its way to help eligible voters register for the 2020 election. They were attacked by police with pepper spray and tear gas after pausing to pray for eight minutes and forty-three seconds (the amount of time Floyd was initially reported to be held down) to remember Floyd near a downtown monument. As this was the last day to register, they were prevented from getting to the polling site and were unable vote as a result.[40]

In contrast is the response of authorities and police when dominant, white populations protest. The starkest and most troubling example in recent history was the mob attack on the Capitol Building in Washington on January 6, 2021. Following a rally organized by Donald Trump and his team to continue the outrageous lie that the election was "stolen," hundreds of his supporters stormed past the barricades and broke through the windows and doors of the Capitol. There is ubiquitous video evidence of these insurrectionists physically attacking the police with lead pipes, shields, pepper spray, and other chemical irritants.

Not only did they desecrate the Senate and House chambers with violence, but some actively hunted elected officials including Vice President Mike Pence (who refused to follow Trump's demand to decertify the election results), Speaker of the House Nancy Pelosi, a Democrat, and other key politicians who the insurrectionists felt were their enemies. Fortunately, they were unsuccessful, but the video footage is chilling as it showed how close the mob actually got to their intended victims.

The police for this event were outrageously underprepared, so much so that it looked suspicious. Prior to the attack, officials had full knowledge that on January 6—the official date for Congress and the Senate to certify election results—30,000 angry Trump supporters were expected to arrive at a rally organized by the president himself, who had whipped up their anger by lying for months about the "stolen" election, saying that he had won it by a landslide, not Joe Biden. An FBI report was shared to all law enforcement agencies the day before, January 5, warning of a possible "war" by right-wing extremists. There was widespread, easily accessible chatter found on alt-right, white nationalist social media channels in the weeks beforehand that openly discussed violence and insurrection in a bid to overturn the election results.[41] Yet, unlike at the BLM protests a half a year earlier, only the Capitol Police were initially present, and they were mostly dressed in soft uniforms rather than hard riot gear. There were no Secret Service, National Guard, local police forces, or any other federal agencies involved as there had been six months earlier at the BLM protests. The results of this infamous day that exposed the vulnerability of U.S. democracy included the deaths of five people including a police officer, dozens of police officers seriously injured, and Donald Trump being impeached for a second time in his short tenure, a first for any president in history.

In watching the footage, it's easy to see the graphic difference between the police and government response to the racially different protest groups—who got the coddle and who got the crackdown. Recall the blending of neurology and sociology: empathy for those most like the decision makers, threat response for those who are different. This, too, is another recognizable pattern of racism: there is a long history of excessive force being used against Black and brown populations who stand up for their rights in the U.S., but also Canada and the U.K.[42]

Inner Skill 5: Self-Education

The inner skill to decode the patterns of power, privilege, and marginalization is self-education. Self-education is really shorthand for taking personal responsibility to understand what the impact of systemic discrimination means for ourselves and our communities—who's doing well, who's struggling, and why.

Like it or not, our social identities matter. Period. Decades of reports and information are available about systemic discrimination in Canada, the United States, and the United Kingdom.

Although our personal experience is a valuable source of information, our pattern recognition system is limited by this very same experience. To expand our cognitive awareness requires effort to learn the stories and histories of nondominant groups, of Black, Indigenous, and people of color while unlearning supremacy thinking that reinforces white dominance.

However, confronting such information is not simply an intellectual exercise. It usually evokes a wide variety of internal responses and emotions. Some may be helpful (such as humility, curiosity, and motivation) and others less so (defensiveness, guilt, rage). Therefore, the previously mentioned inner skills, including self-awareness, self-regulation, and empathy, are required to navigate terrain that, at first glance, may seem very cerebral in nature.

Self-education to help expand both our racial and psychological pattern recognition tool kits may take a variety of forms, including the following:

- **Self-study:** Informally and on our own time, digging deeper into the issues of race and systemic discrimination, to better understand ourselves and our identity groups in the context of the greater whole. A wide variety of resources on this subject is available online, in books and journals, and in mainstream and alternative news sources.

- **Learning through relationship:** Immersing ourselves in environments and activities that expose us to people of different ethnocultural identities to build personal connections and two-way relationships. There is no limit to where or how this happens: employee resource groups at work, events at community centers, sports or arts organizations, or even in connections with neighbors we know less well. The online world can also help bridge physical distances, making a variety of social connections possible.

- **Formal study:** Taking courses through learning institutions such as colleges or universities.

Overwhelmed Yet?

There is more to be said about self-education and inner tools, but I have to be honest. Even as I write this chapter, I feel a little weighed down.

When I get into a structural analysis of power, two emotions usually compete for my attention. I feel empowered and, at the same time, overwhelmed.

I find the lens offered by Sidanius and Pratto empowering because it helps me see things I couldn't see before. It's helpful because my understanding of the issues deepens, fueling my compassion and my desire to make the world a better place.

I also encounter feelings of heaviness, realizing how big the problems of racism, discrimination, and oppression are. This fuels my anger and helplessness. Part of me just wants to go back to bed.

I struggled to write this chapter on power, because it represents an inner tension between my past and present self. In the next chapter, I'll share how I spent many years using this lens on power, got depressed by the negative perspective on the world, and burned out. I'll also share tools I found that helped me tap into inner resources to maintain a hopeful yet complex outlook, so I could still wake up in the morning and take on the day.

Power Part 2: This Time It's Personal

Limitations of a Power Analysis

Earlier in my career, I defined myself as an anti-racist activist and believed deeply in a structural power analysis similar to that presented in social dominance theory. This analysis was the lens through which I viewed the world. It is what I was drawn to, learned, and then taught to others. It's a powerful tool that helped me recognize racial and oppression patterns that I couldn't see before. Prior to encountering theories of social dominance, I had not considered the distribution of resources in society, nor how decisions were made and by whom. Once learned, the systemic patterns became so obvious I couldn't *un*see them, similar to the way we can't help but read the text on public billboards or posters if we are fluent in reading—it becomes automatic. And those who don't integrate racial power into their pattern recognition system may fail to take into account critical information that's necessary to enhance fairness and justice in the world.

Social dominance theory, anti-racism, and other similar theories examine the structure of power, predominantly from a socioeconomic, historical, and political perspective. These critical lenses provide a bird's-eye or system view that reveals patterns of bias, discrimination, and privilege that otherwise may be hard to notice. This perspective on the society better reflects *the experiences and perceptions of nondominant group members*, and therefore challenges the status quo. Looking at power dynamics from this angle may help prevent racial interactions from always being reduced to the level of the individual.

Meaning, when our pattern recognition systems are attuned to social power dynamics, it may be easier to notice the presence of subtle sexism, such as women being interrupted more often than men in workplace meetings. Or it may encourage a manager to take a second look if they notice that a Black employee seems "unable to advance" within the organization and realize that this may be evidence of system-wide racial bias rather than an accurate reflection of the employee's abilities.

A power analysis also helped me make sense of my personal experience of feeling left out and of growing up wanting to be white. Understanding that such experiences were common to other people brought me relief; what happened to me wasn't my fault. It wasn't about my deficiencies as a person, but rather deficiencies in the culture, in institutions, and in how we've been socialized.

This perspective, however, became the exclusive lens through which I viewed the world, and it contributed to me burning out within a half dozen years of work as an activist.

As much as it's important to understand how systemic power operates, it can also be a limiting framework. Given the complexities of organizations and of society as a whole, such theories of power cannot adequately capture all the nuances of our lives.

For example, in a profession that's dominated by women, such as nursing, can a man not experience ridicule or occasionally feel marginalized because of being in an unconventional role? Similarly, in workplaces where racial minorities are in the majority and are in positions of authority, is it not possible for a white person to struggle to find acceptance and a sense of belonging? And what about the success of nondominant group members like President Obama who have surpassed the "sticky floor," broken through the "glass ceiling," and achieved the highest levels of success? How do we name, acknowledge, and honor such experiences?

Although the goal of a structural power analysis is to assist in liberation and emancipation, it can, like a double-edged sword, cut both ways and fuel a sense of despair. This is what happened to me, and it was a very common experience among my social justice peers.

We valued "critical thinking," but that meant offering critiques exclusively from the left side of the political spectrum. Avoid purchasing coffee, books, and clothing from large corporations. The problems in the world are created by straight, white, able-bodied men. Conservative views are backward and racist. The Olympics are a waste of money and resources.

We were outside the box. Alternative. Organic. Vegetarian. "Woke," using today's language. We were on the side of justice and we held the moral high ground (at least from our own viewpoint). We brought our "critical perspective" everywhere we went. At dinner parties, any statement could be turned into a political moment to "educate" or "call out" others. It could be a comment about the food we were eating (*toxic hormones in meats and veggies*) or dishwashing soap (*phosphates poison marine life*), or critiques of mainstream politics (*all big political parties are on the right side of the spectrum*). We could critique and therefore rain on anyone's parade, including our own.

An acquaintance who strongly held feminist and anti-racist perspectives once told me she found social gatherings stressful. I was no introvert, but I could relate to what she said. I frequently had to brace myself in informal group settings, feeling an invisible pressure to somehow step in and respond to people's "ill-informed" perspectives about how the world worked. I was on guard in most contexts. I dreaded the possibility that someone might say the "wrong" thing and I'd have to "correct" them—exhausting!

I didn't realize what was happening, but pessimism and disapproval were becoming my close companions. And there was never time to rest or celebrate successes as perfectionism was a bullying internal presence that was fixated on trying to fix the brutality of the world; therefore nothing we or others did was ever "good enough." Unaware that the dial on my negativity bias was stuck on high, I'd lost the ability to distinguish critical from cynical. After all, we were always surrounded by some form of oppression or another—racism, sexism, heterosexism, ableism—with institutions and corporations just waiting to co-opt another good idea "we" came up with and monetize it. It was a bit of a bunker mentality in the trenches of social activism. We were a minority and power was not in our favor. We wanted the world to be beautiful, but our words and actions communicated that it was mostly ugly.

As it's said: what we look for, we find. When we view the world exclusively through the lens of power as defined by oppression-based theories, we can feel paralyzed by perfectionism and fear of judgment by our peers who are also lost in the same spin cycle. There is a danger of our perspective becoming negative and despairing. For me, life looked pretty gray. The critical lens I had adopted eventually helped me burn out. It wasn't the only factor, but it gets the nod for Best Supporting Actor in a Comedy That Is My Life.

Reflecting on this now, I've come to realize that the ways we think about things—the lenses we use—also become neurologically wired and formed into habits. When the same lens is used regularly, for example, in social activism and academics, we can almost lose choice in how we view the world. Just as privilege often begets privilege, the opposite can also happen. If we are immersed in issues of marginalization, it's hard not to see marginalization everywhere. Unless we are aware of this particular psychological trap, we may, on the way to empowerment, inadvertently enhance our sense of victimization. They are, after all, flip sides of the same coin.

Such perspectives on power also helped neatly divide the world into three specific groups: victims, perpetrators, and rescuers (us). A key subcategory divided the people into those who "got it" or were "woke," and those who were not. Ironically, Us/Them was alive and kicking, being perpetuated by my social justice peers and me. We just didn't recognize it. And not surprisingly, no matter how far up on the moral high ground we are, the view is still cloudy.

Personal and Social Power: A More Balanced Approach

Although I have become critical of my past orientation, I still believe that to enhance democracy and nurture a stronger sense of inclusion, a systemic analysis of power is essential. That's why the entire previous chapter was dedicated to social dominance theory in order to support our pattern recognition of oppression dynamics. It *is* important to understand how power, institutions, and structures influence our choices, interactions, and culture both overtly and covertly. Fundamentally, the goal of such a power analysis is to help alter our awareness of social norms, so that we can notice unjust patterns in order to bring positive change at individual and systemic levels. And that's something I firmly support.

I do take issue, though, with it being the only acceptable perspective on the world. We're not just cogs in a wheel. We're influenced by and influence our environments, including our families, friends, organizations, and society at large.

As my discomfort with this single perspective grew, I began to explore other worldviews. I discovered a more nuanced, balanced approach to power from the Process Work Institute. Pioneered by Jungian psychologist Arnold Mindell in the 1970s, this community has been developing expertise in understanding individual and group behavior, especially related to intergroup conflict. According to Mindell, we have access to both *social power* and *personal power*.[1]

We receive social power from two sources. The first is status gained from our various social identities, including race, gender, class, sexual orientation, and ability. Called *global rank*, it's the level of power we have (or are given) when we walk into a room full of strangers; it's the first impressions, both conscious and unconscious, based on what we look like and how we speak, hold, or present ourselves. We don't have a lot of control over most of these factors. They are a result of the body and circumstances we were born into.

The second source of social power, gained from our status in our immediate environment, is our *local rank*, and it may shift depending on our microcontext. For example, it may be very high at work because of our executive job title, but when we go shopping for

	Rank and Power Model
1. SOCIAL POWER	**Global rank:** Context dependent, seemingly static, associated with social norms, random (e.g., race, wealth/class, religion, gender, sexual orientation, physical ability). **Local rank:** Context dependent, shifts rapidly, associated with local norms, values, current conditions, participants, topics (e.g., seniority or position in group, adherence to norms, popularity/insiderness, communication style).
2. PERSONAL POWER	**Psychological rank:** Life experience, emotional fluidity, communication skills, humor, relational skills, insight into self and others. **Spiritual rank:** Connection to larger purpose, vision, transcendent experience, knowledge of self and others, awareness of death and life.
Source: Julie Diamond, "Deep Democracy: Creating Whole Systems Change," workshop held by Anima Leadership, Toronto, October 2010, www.diamondleadership.com; training based on the work of Arnold Mindell, Process Work Institute.	

groceries, the part-time clerk has more power over decisions than we do. Such status can shift from moment to moment depending on the context and the people we are around.

A key point to understand is that, at times, our local rank may temporarily outdo our global rank. For example, I've been in environments in which women of color, who may have lower global rank based on identity, possessed the highest levels of local rank—and therefore influence—within their organization. This was due to their expertise, seniority, and ability to articulate the issues. Equally I've watched white men—high global rank—have difficulty participating and speaking up in conversations on racism or sexism. Their inexperience with issues in that specific microcontext held them back.

Social power is the link between Mindell's model and social dominance theory (or my early training in anti-racism), as both seek to identify the different social relationships between groups. Mindell's model acknowledges that our groups—race, gender, class, sexual orientation, ability, and so on—are important sources of power. But they can also undermine us.

High status from our social identities (social power) is considered the most brittle type of rank. It comes largely from outside ourselves. We didn't have to do anything to gain entry, so it's an invisible privilege. In some cases, higher rank because of something like our race or sex (being white or male) may actually be disempowering. First, because we may not be aware of it, and second, because it may not make us *feel* powerful. Further, this invisible quality almost always leaves us vulnerable to making mistakes and being criticized, ridiculed, and even attacked based on our unearned privilege (and lack of awareness of it).

The resulting emotional state is one of "fragility," often resulting in defensiveness, anger, tears, or denial when confronted with the realities of dominance and oppression. As explained by white anti-racist educator Robin DiAngelo, who coined the term *white fragility*, many situations can "trigger racial stress for white people," resulting in unhelpful emotional reactions to regain their "racial position and equilibrium."[2]

The real clout behind Mindell's model comes from the next category, called personal power. This is something we all possess. Such power was drawn on by heroes of history such as Mahatma Gandhi and Dr. Martin Luther King Jr. Arguably, these leaders had lower global rank in their contexts. Of course, each of them had some social rank to begin with; they both were middle-class, university-educated men. But in their time periods, race was the defining social identity related to the legacies of slavery, segregation, and colonization. Nonwhite people were overtly classified as inferior and less human, and, accordingly, were treated with contempt and violence. Gandhi, a lawyer by training, was still a soft-spoken brown man fighting for civil rights for his people in South Africa and India against the all-powerful British Empire. King, a church leader, was a Black man living through the intensely racist pre–civil rights era United States. Both of these individuals, however, possessed inner strength and conviction, power that flowed from their psychological and spiritual qualities.

Psychological rank is status from our personal attributes—who we are and what makes us unique. These attributes include our life experiences, leadership qualities, sense of humor, self-awareness, communication skills, courage, curiosity, and community consciousness.

Personal power also comes from *spiritual rank*. This form of influence comes from the ability to make meaning of things and place ourselves within the larger framework of life. Part of this process is struggling with existential questions: *What's my purpose? Why are we here? How do I deal with life's tragedies and hardships?*

A deep knowledge of self and others and an awareness of the cycle of life and death are also markers of spiritual strength. This type of rank may or may not flow from organized religion. But usually, it is accompanied by the ability to celebrate the beauty of life while also acknowledging its brutality.

Gandhi and King both exemplified such personal power at profound levels, as did many of their peers. In the face of brutal oppression based on race—whether it was British rule in India or racial segregation in the United States—these leaders did not let their lower global rank hold them back. Their power—based on personal convictions, interpersonal skills, and deep spiritual callings—was sufficient to inspire millions to unlock their own personal power, tackle racial oppression, and create powerful social change.

High rank in the personal power dimension fundamentally boils down to a *feeling* of empowerment—of being able to do, to advance, to risk—even in the face of opposition, uncertainty, and low odds. And it is not just the domain of celebrated heroes of history. It's Javier Espinoza from California, whom we met in Chapter 5. He overcame a childhood of violence to work with women and children who suffer abuse. It's the white hockey coach in the Toronto suburbs who pulled his team from the ice because of racial slurs used by the opposing team.[3] It's the Black grandfather who carried an injured white nationalist to safety when far-right counter-protestors clashed with Black Lives Matter supporters in London, England.[4] It's the child in the schoolyard who is willing to intervene when they witness a peer being bullied.

So the good news is that we all possess personal power. The better news is that personal power can be grown and developed.

This is the most hopeful aspect of Mindell's model. We can always cultivate and increase our feeling of personal power. It can be enhanced through inner work, through developing our psychological and spiritual capacities. Personal power can be grown through any process that helps us improve our sense of who we are and how we interact with others. It helps us find purpose, meaning, and joy in our lives. Because this form of power is internal, it is more resilient than social power. And it can be applied across different contexts.

Further, changing the dynamics of social power, especially around issues of race, is a long-term project. Marginalized communities may see little immediate change to help their circumstances. Personal power, though, is more direct. Regardless of where we are in the social hierarchy, we can rely on personal power every day to make our experience of living more positive, or at least more bearable if our circumstances are dire. Developing personal power isn't anything new. We harness it whenever we attempt to make a small or big change to improve our lives. It's what support workers at a women's shelter draw on to do the work they do. Personal power is also accessed when victims of violence find ways to heal and move forward with their lives. It's at the core of peer-led leadership networks designed to help youth launch community innovation projects.

Any meaningful social change project, in fact, is about empowerment of people—about agency rather than dependency.

In the Deep Diversity framework, personal power and the potential for change it represents can serve as a catalyst for system-wide change. There is always a dialogue, a creative tension, between recognizing the influence of institutions on the individual and of the individual on the system. The trick is holding this tension and not getting caught in either-or thinking. This is the balance point, both emotionally and intellectually.

How Characters from the Social Justice Universe Misuse Power

With greater personal power we can develop the inner skill of *discernment*, the ability to respond thoughtfully rather than simply react emotionally to a situation. Although we'll discuss this concept in more detail shortly, it's important to introduce here. In my consulting practice, I fairly regularly encounter leaders who misuse or abuse power because they fail to *discern* correctly what's happening in front of them with regard to systemic forms of discrimination.

To understand this more deeply, I'll introduce three different archetypes from the social justice universe—Paper Tiger, Free Speech Centrist, and Evangelical Activist—to explain how a lack of discernment and low personal power may function as barriers to achieving justice goals.

Take, for example, Thomas, a composite of a white executive in a large public sector organization. Because of his inability to recognize how patterns of subtle racism play out in society broadly or within his institution specifically, he became defensive and argumentative when allegations surfaced of systemic racial bias in hiring, promotion, and pay equity in a report about his employees. Rather than listen to the experiences and concerns of the employees of color, he responded in very common ways: called those individuals who raised the concerns "troublemakers"; blamed poor decision-making on factors outside his control; and declared he didn't have "a racist bone in his body," citing the letter of support the organization had made in support of Black Lives Matter. In short, he tilted away from the problem and refused to take any real responsibility for it, a terrible misuse of power. His emotional fragility was on full display in his responses: minimizing, denying, and individualizing the problem while pathologizing the employees who raised concerns. Stuck in an unwavering belief in meritocracy—that people succeed based on hard work—he lacked the ability to discern how racist patterns were operating under his leadership and was not open to understanding there was something he might be missing.

Unfortunately, there are too many leaders like Thomas who misuse their power because of their inability to decode oppression patterns and become systems thinkers regarding issues of diversity and justice. Just as the struggling canaries in coal mines were harbingers of toxic air for workers, racial bias complaints in organizations are often omens of systemic problems that, if left unaddressed, may poison the environment.

I refer to such leaders as *Paper Tigers* as they suffer from an excess of ego and entitlement including internalized dominance and fragility tendencies. They have been socialized into unbalanced ways of being, including always knowing the answer (even when they don't), binary either-or thinking, being the center of attention, and sticking compulsively to the status quo while struggling with vulnerability, authentic self-reflection, and perspective taking. It's amazing the disproportionately high-profile leaders I've met who turn out to be Paper Tigers whether in organizational, academic, political, psychological, scientific, journalistic, athletic, or organizational contexts.

This absence of pattern recognition is also the problem within many self-described liberal-minded people, what I call the *Free Speech Centrists*. For example, in an online video conversation, author Irshad Manji and Debra Mashek from the Heterodox Academy were lamenting the political polarization in the United States. Specifically, they were criticizing "callout culture"—the tendency for justice-minded individuals to criticize and publicly shame those who have made racial or identity-based mistakes, acting inappropriately or harming others in some significant way. As proponents of dialogue and free speech, they made many important points, including how knee-jerk callouts can create environments where people don't speak their minds for fear of being ostracized.[5] I am in agreement with their central proposition that *how* we achieve social change goals is as important as what we achieve because process matters; that we need to engage with those we disagree with, and especially so during polarized times; and that activists can sometimes behave like those that they oppose.

Their discussion was minimal, however, about the problem of racial oppression itself. They spent little time on the problem of racism and focused primarily on how unhelpful racial justice activists were being because of the methods used. In this view, the *reaction* to the racial violence seems more important than the violence itself. Their inability, or unwillingness, to meaningfully explore the patterns of systemic discrimination left the impression that the problem of racism was really an individual thing that could be fixed if we just talked nicely to each other.

This is common among Free Speech Centrists as they emphasize dialogue and speech but weaken their arguments when they don't articulate the system-wide devastation related to racism. The words and ideas they choose to emphasize demonstrate a lack of discernment regarding patterns of oppression.

Furthermore, there is a reluctance to actually acknowledge that social change is complex, unpredictable, messy—that there is no formula. Civil rights legislation in the U.S., for example, was passed in 1964 not because of polite intergroup dialogue between white and Black people, but because of a variety of nonlinear methods including mass mobilization, nonviolent direct action, confrontations, boycotts, and, yes, shaming. Even violence played a critical role in changing public opinion especially when people saw on television how peaceful Black protestors and their allies were attacked by police and other state agencies with attack dogs, batons, shields, fire hoses, and bullets.

Discernment, however, is also missing from a particular brand of social justice warrior, which results in a different misuse of power.

I call them the *Evangelical Activist*, an individual who brings searing clarity, skill, and assertiveness in identifying the dynamics of power, privilege, and oppression in any situation. They are articulate and fearless in "calling out" racial or gender bias in interactions in the context of community or the workplace with peers, direct reports, or even bosses. However, the shadow side is that they can be judgmental, intimidating, and punishing in their use of shame and blame tactics. Coupled with high levels of distrust, they sometimes behave and talk as though there is a singular path to "ending oppression" and they are among the chosen few to shepherd the illiterate masses along.

In a particular organization I worked with, one such individual had ground a community project to a halt. I was brought in to do some work with the group and watched with fascination as an Evangelical Activist maneuvered and intimidated their peers into submission. During team discussions, many people looked down or literally had their heads on their desks, terrified to disagree in any way as the Gospel of Anti-Oppression was shared. The only people who spoke were very deferential, offering atonement for their past racial or identity-based transgressions. At the end of the session, the Evangelical Activist offered public compliments to those who had "most improved," like a teacher handing out gold stars to children.

I've seen this particular pattern enough times—within myself and my peers—to know that the power of the Evangelical Activist often comes from some combination of the following:

- academic/activist training that promotes an overreliance on perspectives that divide the world into victims, perpetrators, and rescuers
- academic/activist training that nurtures an unwavering moral certainty and self-righteousness, with an encouragement to use punitive justice methods to intervene and interrupt whenever they believe oppression is happening
- personal wounding or identity-based trauma that guards unprocessed emotions such as fear, grief, or vengeance

As a result, Evangelical Activists overidentify with their low social rank and are often unable to discern how much personal power they actually wield. Because of belonging to a minoritized group based on factors like race or gender, they believe they are immune to committing abuses of power—because from their worldview, they don't possess any. And this is how the misuse of power arises, because they often underestimate how much their words, actions, or distrust impacts others from both dominant and non-dominant groups.

They lack discernment skills, overusing the language of "trauma" and "harm," often categorizing small unconscious slights and bias missteps in the same bin as hate crimes. As a result, allies and antagonists become indistinguishable to the Evangelical Activist as many relationships around them are brittle and precarious. Sometimes an unconscious desire for revenge—to have others experience some of their pain and discomfort—is hidden underneath disproportionate demands of accountability from others. They don't realize how their words and actions promote greater polarization, fear, and rigidity, ultimately becoming antithetical to a transformative justice approach where reconciliation, reparations, and being our whole selves is more possible.

It's important to note that while the three archetypes exist along a continuum, they are not proportional in numbers or impact. Paper Tigers are the most numerous and the greatest block to meaningful change. Toward the middle are Free Speech Centrists, who would be powerful allies if they could develop more expansive racial pattern recognition skills. On the other end of this asymmetrical dynamic are social justice activists, a tiny minority compared to the systems they hope to change. For context, Evangelical Activists are a minority within this minority. The reason I am drawing attention to this group is that they are often leaders in social activism, influencing the tenor, shape, and direction of how change takes place in the long and short term. As a result, they are a group with outsize impact.

Applying Personal Power in the Social Justice Universe

I'd also like to suggest that these characters from the social justice universe—Paper Tigers, Free Speech Centrists, Evangelical Activists—are often within each of us, depending on the moment and context we are in. And they are only a handful of voices within a sea of perspectives that need to be heard in order to achieve long-lasting, transformative racial justice goals.

Although entitled and egotistical, Paper Tigers partially echo the views of many white people who are confused about issues of race but are otherwise well-intentioned and fair-minded. For example, one white participant approached me partway through a diversity training session and asked: "Are you going to deal with what white people can do? I'm afraid of saying the wrong thing, of making a mistake, and being told I've done it wrong." She was truly distressed, vulnerable, and confused about how, as a white person, to move forward on racial inclusion.

If we are from the dominant racial group, personal power is necessary to help us accept that, in countless ways, our group benefits from being part of the social norm. There is systemic bias, and it plays in our group's favor, requiring extra effort to discern the patterns that are more obvious to nondominant groups. Individuals need psychological and spiritual strength to deal with the emotions that surface with that discovery, and with the inevitable mistakes and criticism that come with possessing social privilege. And personal power is required to listen to the pain experienced by minoritized groups and individuals, rather than dismissing or minimizing their perspectives because they make us feel bad or accused. Although we may not have created the problems of racial injustice, we benefit from it and have responsibility to make things better.

On the opposite end is the perspective of the Evangelical Activist. In spite of their judgmental, Leftier-than-thou attitude, they reflect the perspectives and experiences of many nonwhite people who struggle with how hard it is to have their racialized experiences acknowledged by mainstream society. I'm reminded of the pain and vulnerability of living on the wrong side of privilege expressed by a

Black mother during a training session. She wept with frustration and anger as she asked me: "Why is my son suffering the same racial taunts and humiliating schoolyard experiences that I did forty years ago? Why is this still happening?!"

If we are from nondominant racial groups, personal power is required because, with disproportionate regularity, the stakes shift from ostracism to life and death. When little boys or young men from our community are killed in broad daylight because of their skin color while the police officers who shoot them are let off the hook, there is a place not just for anger but also for rage, for irrationality. To deny such righteous anger would be to deny our humanity.

Personal power is needed by minoritized people to find healthy ways to process pain and loss, to make meaning and not sink into cycles of vengeance, despair, or self-sabotage. When we do so, our experiences of struggle can make us stronger in both small and significant ways. To recognize that we can see what dominant groups cannot about themselves and the world around them is also a form of transformational power. Change, after all, comes from the margins. And it is the marginalized who hold the keys to the future, not those who desperately cling to the status quo.

Furthermore, for social justice activists it's important to remember what was discussed in Chapter 2, that the pain of social exclusion evokes deep, unconscious survival responses within all of us; a desire to avoid death is wired into our neural architecture. All humans are built to survive and our groups will do whatever it takes to live regardless of our high or low status. So, as challenging as the work already is, it's imperative to develop social change strategies that not only work to bring minoritized groups toward the center, but also keep dominant group members within the circle of "we." Which is why an overreliance on tactics such as shame, blame, and ostracism to make dominant group members "change"—whether white, male, or straight—often backfires. Put another way, such strategies may activate an unconscious impending-death sensation that people will do anything to not experience. Knowing and working with such complexity, I believe, is critical to moving toward transformative social change.

Inner Skill 6: Discernment

Rhonda Magee, a renowned teacher who shows learners how to use mindfulness practice to strengthen their racial justice muscles, describes the key inner skill of *discernment* as the ability to make the "important distinction between automatic judgment and [a] more considered, evaluation."[6]

If we are from nondominant or minority groups, we need discernment skills and personal power to not be overwhelmed at a society that is imbalanced against our group, and not give in to despair over the magnitude of the problems we face. Discernment and greater rank in psychological and spiritual dimensions would be necessary when our inner Evangelical Activist emerges, to see the world as a place of opportunities and beauty in order to navigate the roadblocks with hope and heart, not despair or self-righteousness. Or when the Paper Tiger within us becomes reactive and ignorant to what's happening in our context. Rather than bulldozing ahead, can we slow down enough to listen to others, to access empathy in order to feel a connection to experiences we don't understand?

We all need discerning personal power to accept that changing social norms is a slow but continuous process, usually beyond the scope of a single lifetime. It's also necessary to resist the temptation to dehumanize any group—whether high status or low—to not see individuals as two-dimensional, cartoon symbols of a community.

Inner Skills 7 and 8: Relationship Management and Conflict Competence

From my experience, *relationship management* is essential to negotiating power and its unequal impacts on social groups. Relationship management is more like a skill group than a single aptitude. It includes the ability to build trust, motivate, negotiate, and work well with others, and deal effectively with disagreements.[7] It also builds on the inner skills of self-awareness, self-regulation, and empathy, supported by compassion and discernment.

Among the abilities that fall under relationship management is *conflict competence*, which is important enough that it is an inner skill in its own right. According to Julie Diamond, one of the cofounders of the Process Work Institute, "the skill of negotiating difference, at the heart of it, is a conflict skill because we have to navigate through different interests, positions, and stakes."[8]

Conflict competence relies heavily on personal power. It includes listening to another person's perspective and letting it affect you. It's about speaking honestly, vulnerably, and truthfully, which means standing up for yourself and your side in a conversation.

It doesn't mean that each conversation must end up in a mushy middle ground where both sides feel they have to "compromise" important things. An authentic conversation and willingness to step into the heat of conflict can be transformational, taking us to a new understanding or a greater depth of empathy. Ultimately, we may arrive at a new place in a relationship from which we might make decisions. It can also help us see what's unique about another person, diminishing our tendency to use stereotypes of out-group members.

Diversity, at its core, is another word for difference. To negotiate difference we have to work though tensions, whether at the group or interpersonal level. And we also have to negotiate significant internal conflict, mental states in which we struggle consciously or unconsciously between incompatible needs, drives, impulses, or desires in relation to our in-groups and out-groups.

Diamond expands on this point in a practical way, highlighting the importance of making mistakes:

> There's a lot of uncertainty and difference in diversity; there's no way to be perfect at it. At some point you will make mistakes and you will be confronted by others. You are going to use the wrong term, you are unknowingly going to insult somebody and get corrected for doing so. You are going to be shamed—and for some people, this is really hard. They can never be wrong, there's too much shame, and they shut down. Completely. One of the greatest skills is the ability to publicly say, "Oops, I got that wrong."[9]

What would it be like if, when we were feeling polarized or distressed, we could see even a little of the experience from the other person's perspective? If we could set our political views aside enough to be a little curious about the other person? What if they could offer their strong perspectives, while also validating ours?

From my experience, conversations are potentially transformative when people in polarized positions are able to be empathetic to each other. There needs to be a give-and-take in the process, especially acknowledging the other person's perspective. It also requires pattern recognition skills that allow us to discern how one's social identity may be playing a role in the interaction.

Conflict competence can serve to strengthen relationships between people and groups, and thereby nurture healing. The key is for conflict to be handled well.

An Unexpected Response to Terrorism

I learned this lesson firsthand in a rather dramatic way. Years ago, I was working in the Netherlands with young leaders on issues of racism and cultural integration. Our team had been invited to create a leadership program for the Dutch context, to tackle the thorny issues of multiculturalism, immigration, and belonging in a post-9/11 Europe. This intercultural dialogue took place in a retreat setting outside Amsterdam.

Nationalism was rising at the time, accompanied by anti-immigrant, anti-Muslim sentiments. The previous year, artist Theo van Gogh had been murdered in the streets of Amsterdam by an Islamic radical for making an inflammatory, anti-Islamic film and calling Muslims "goat-fuckers," among other choice racist slurs.[10] According to opinion polls at the time, Islam, immigration, and integration had been at the top of Dutch minds.

Feelings of Us and Them were intense in a society that lived significantly racially segregated realities. It was in this context that young leaders from across the country, aged seventeen to twenty-four, were invited to attend a two-week residential peace-building program. About 40 percent were racial minorities and 60 percent were white; some participants had self-selected and others were sent by their organizations.

Our program, a precursor to the Deep Diversity framework, supported the participants on the levels of both head and heart. Intellectually, we offered them concepts that helped expand their understanding of race, discrimination, and social integration—to develop pattern recognition skills. We supported their hearts through activities that taught self-awareness, empathy, and the ability to name and express their emotions. There was an emphasis on relationship-building skills, including listening, communicating, and managing their emotional triggers.

We also offered several intergroup dialogue formats, including opportunities for white and racial minority participants to share their experiences of belonging to, respectively, dominant and non-dominant groups. There were ample opportunities to talk about their own experiences—perspective giving—and to listen to the experiences of others—perspective taking.

The conversations sometimes grew heated, eliciting painful experiences and exposing stark political divisions. Feelings of shame, anger, guilt, and pride were strongly expressed on both sides of the racial divide. This was our first year designing this training process, and in spite of significant facilitation experience on our organizing team, we wondered at times if we had let it get off track.

The true test of the program, however, came in an unexpected manner. On our second-last day, we had just finished our Community Day. The young leaders had showcased their learning by teaching some key concepts to friends, family, community, and political officials. The event had gone extremely well. Participants demonstrated their newfound perspectives and camaraderie. They conducted themselves with grace and elegance, even in the presence of an inflammatory high-level politician whose caustic words promoted intolerance toward Dutch people who were Black or Muslim. Our youth were on a high, and I felt very proud of them.

The date, however, was July 7, 2005. Just as Community Day was wrapping up, we received breaking news that the London Underground, less than an hour's flight from where we were, had been bombed. The world would later learn that four homegrown terrorists had targeted the busy transit system with suicide bombs, killing fifty-two civilians along with themselves and injuring about seven hundred people. At the moment the news broke, we knew few details: only the destruction, smoke, death, and suspicions of a terrorist attack.

Our staff team struggled with what to do. We had limited media access in that remote location, and most of the participants did not even know the event had occurred.

We were worried that sharing this news could derail everything we'd worked toward over the last two weeks. It could kill the spirit and, likely, split our community on the day before we were to depart—a terrible way to end the program. Our team, however, felt there was little choice. We had to disclose to the group what had happened and let the cards fall where they may.

We gathered all seventy youth leaders, and I apprehensively shared the news about London. Not knowing how else to proceed, we simply opened the floor to comments.

Rehana, a Moroccan Muslim, despaired: "9/11—it's happening again." Everyone in the room implicitly understood her fear of an anti-Islam backlash. As she wept, others nearby placed their hands on her, offering comfort. A young white woman, Ingrid, shared that her sister was in London and was not responding to calls to her cell phone. Fearing the worst, Ingrid raced out of the room in tears. Remarkably, she was followed by Alisha, a young Black woman, who went to comfort her. I say remarkably because these two women had been at loggerheads for the entire program, occupying opposing views on how racial minorities should or should not "integrate" into Dutch society. I thought they hated each other.

Abdul-Hamid, a former child soldier from East Africa, was completely distressed. His full scholarship to an Ivy League school was in jeopardy, he shared, as he now had a poor chance of obtaining a visa for the United States. Abdul-Hamid's goal was to use his education to help rebuild his war-torn country and support family members who still lived there. He put his head in his hands, trying to push back the tears. A large, lumbering white guy, Auke, protectively placed an arm around Abdul-Hamid's shoulders, offering silent support.

And so it went on spontaneously, without any intervention from the staff team. Participants took turns speaking from the heart, listening to each other, weeping, and offering each other what comfort they could.

It was the most remarkable and generous collective act of compassion and healing I've ever been a part of. As I walked around that evening, in many corridors I saw lit candles and small, diverse clusters of participants talking quietly, holding and supporting each other.

It took me a while to articulate what I was witnessing. And then the penny dropped: they were *grieving*. There was deep sorrow and it was implicitly understood that "we" had been affected. There was no blame or finger-pointing, except toward radical terrorists. There was no "Us" or "Them"—the loss was collective. This tragic event didn't split us along racial and religious lines, as I had feared. It brought us closer together. This group of participants did what all of our communities should have done, which was to grieve and mourn our collective losses.

A public act of terrorism was countered by a communal act of healing. I was overcome by this experience, by the compassion and deep learning it offered. It couldn't reverse the events, but it dramatically counteracted the reactionary responses heard in much of Europe.

Had this been the second day of the program rather than the second-last day, I'm certain this tragedy would have had a profoundly negative impact on the sense of community. Instead, the program became a testimony to the power of relationships—with the necessary head and heart skills in place at both interpersonal and intergroup levels—to effectively manage conflict and develop resilience.

Questions for Cultivating Relational and Conflict Competence

Many tools, books, training programs, and resources are available for developing relationship management skills and conflict competence. The following questions may serve as helpful starting points to developing pattern awareness:

- Generally, how much attention and time do I spend helping and supporting others? How frequently do I ask, *What can I do to help*?
- How well do I listen? Do I both ask questions and fully attend to the responses? How frequently does my cell phone, tablet, or computer distract me during conversations or meetings?
- How skilled am I at noticing nonverbal communication in others and myself, including tone of voice, facial expressions, eye contact, posture, or gestures?
- How effective is my communication with others when I personally need help or support? Am I as clear as I could be?
- During disagreements or conflict, how skilled am I at standing up for myself and my own perspective? How easily do I let my position and needs go?
- How skilled am I at listening to the other person's perspective? Can I be curious and empathetic, asking about their viewpoint, or do I get more entrenched in my position?
- How frequently do I invite feedback from other people? How well do I receive feedback (that may or may not be invited)?
- How aware am I of my own feelings and needs in relationships? How aware am I during disagreements or difficult situations?
- When things get stressful and heated, how effectively do I manage my own emotions? How quickly can I regain a state of centered, calm clarity?
- When the situation becomes stressful or charged, or there is a clash of perspectives and experiences, how aware am I of the feelings and needs of others?
- Do I tend to avoid situations that are heated or filled with conflict? How willing am I to lean in to conflicts and explore ways of getting to understanding or resolution?
- How effective am I at working through disagreements or conflicting needs?

Review this list of questions again, but this time ask them in relation to social identity (race, gender, sexual orientation, ability, and so on). *How well do I listen when issues of social identity are raised? How skilled am I at listening to the other person's perspective regarding race or gender issues?* Relational skills are significant markers of personal power. To transform conflict, we need to step beyond our perspective, and at times consider the issue from outside of our social identity.

What new learnings from these last two chapters on power help expand your racial pattern recognition repertoire regarding systemic issues? Which help further develop your psychological pattern recognition schemas?

Deep Diversity: Bringing It All Together

SOMETIMES WE face situations where conventional dialogue and conflict resolution skills cannot be applied. The impacts of intergroup conflict can sometimes be so painful, personal, direct, historical, or complicated that there is no clear resolution. In such cases, the best we can do is struggle to make sense of our lives and situations in some constructive manner. *Meaning-making*, as we will see in this final chapter, is the inner skill that can help us move forward when nothing else can.

Not Too Late

I was invited to present at a youth leadership conference in northern Ontario, where more than half of the participants were Indigenous. After the presentation, I had planned to attend the screening of a video about Grassy Narrows First Nation, created by a youth group that was getting some buzz at the event. But I chose instead to continue a conversation with Walter, a teacher from Grassy Narrows with whom I was seated at lunch.

I'm glad I did. Walter spoke with great love and affection about his community. Yet, the history of Grassy Narrows—known as Asubpeeschoseewagong in the community's Anishinaabe language—is filled with the pain of anti-Indigenous racism and the destructive legacy of colonization.[1] Public record shows that government policy throughout the twentieth century, for example, forcibly took children from their families to live in the now infamous church-run residential schools—known as industrial schools in the U.S.—which worked to strip away the culture, language, and traditional ways of Indigenous Peoples. As we know today, many of these institutions were rife with physical, emotional, and sexual abuse. Many people were damaged and some died, yet the community of Grassy Narrows survived.

In the 1960s, the people of Grassy were forced to relocate from their original territory—from their traditional subsistence lifestyle that was rich with fishing, hunting, and trapping—to make way for a hydroelectric project. The project destroyed a major food staple, wild rice beds, drowned out the fur-bearing animals that were central to the group's microeconomy, and flooded sacred burial grounds. They were forced to live on a reserve on a stagnant lake with poor soil that supports neither gardens nor their traditional ways. Yet, again, the people of Grassy adapted and survived, and continue to live there today.

In the 1970s, it was revealed that the fish in the lake were contaminated with mercury because of a paper mill upriver. The local fishing economy was destroyed overnight, with 90 percent employment dropping to 10 percent. Over a hundred deaths were recorded as a result of the mercury poisoning, called Minamata disease. (In 2010, scientists found that close to 60 percent of the people of Grassy still have mercury-related ailments, including limited movement of limbs, loss of balance, hearing loss, insomnia, headaches, and fatigue. This includes people who were born long after the mercury dumping had ended.)[2]

Walter told the story of his community taking a downhill spiral as a result of choices, most of which were not their own. In such a context of neglect and poverty, it's not hard to understand why drug abuse, suicide, and various forms of physical and sexual violence took root. Walter also described how the community resisted and survived. He credited the women within this remarkable population for saving them through efforts to combat addictions when other avenues were closed.

To stop the clear-cutting of their treaty lands, Grassy Narrows First Nation mobilized. They began a campaign of road blockades in December of 2002, turning back logging trucks in spite of government and corporate opposition. At the time of writing this book, the protest continues. Grassy Narrows is described as Canada's longest-running First Nations blockade.[3]

Yet what stands out for me most was what Walter said at the end of the conversation about Indigenous Peoples and the rest of Canada: "It's not too late, though."

"Not too late for what?" I asked.

"It's not too late for our relationship—there have been mistakes made on both sides."

Relationship? I struggled to see the "relationship" Walter referred to. Dishonoring of treaties. Forced relocation. Poisoning of the land. Destruction of culture. Abuse, poverty, and neglect. If anyone had reason to be angry and hold a militant position toward others, it was Walter.

His statement floored me. My eyes welled up, as I asked how he found it in his heart to say what he'd said, given the lopsidedness of historical events. He quietly replied, "It's taken a lot of healing, brother, a lot of healing."

I still get goose bumps as I recall the story today. Walter had done something that I didn't understand very well at the time. Although he did not share any details of his healing journey, his ability to speak the words he did demonstrated the presence of powerful inner work that had altered his perspective on the world. He was still fighting the good fight and was still part of the resistance efforts, such as the blockade to protect his community's land and rights. But he spoke of relationship. He spoke of reconciliation under circumstances where many of us might be drowning in despair and hate aimed at those who were the source of our persecution. He had healed enough that even those who might be considered his persecutors—whites, non-Indigenous people—were still worthy of relationship. He was able, under extremely difficult circumstances, to humanize his out-group.

Inner Skill 9: Meaning-Making

Research on resilience reveals that those who are able to make meaning from negative situations in their lives are better able to bounce back from stress, tragedy, and trauma. Many studies with war veterans, many of whom had been brutally tortured, demonstrate this.[4] According to Steven Southwick and Dennis Charney from Yale University, whose expertise is in post-traumatic stress, most of these military personnel had found ways to reinterpret the significance of their extremely harsh ordeal, having "grown stronger, wiser and more resilient as a result of it. They also reported that they were now better able to see possibilities for the future, relate to others and appreciate life."[5]

To make meaning, we need to find constructive life lessons in adversity. Research with people who have suffered loss or trauma—including sexual abuse, rape, or death of a loved one—indicates the importance of being able to reflect on questions such as "Why did this happen?" and "What good can come of it?"[6] Those who wrote about their experiences and used these questions in their reflections seemed to recover and heal more quickly. Compared with peers who did not, they required less health care support over the following year.

Walter had clearly found a constructive way to make meaning of the extremely difficult experiences faced by his community, including overt and subtle forms of anti-Indigenous racism. He spoke directly of having undergone a healing process to reach his state of compassion and forgiveness, and to keep his heart open to relationships with non-Indigenous people.

It's important to note that my meaning-making analysis and Walter's story come with two cautionary points. First, the healing process is a deeply unique, nonlinear process. It depends on many variables, including the availability of community and social supports, access to health care, genetic tendencies, and income and education levels. It's important to recognize that in some people's circumstances, it may be beyond their control—perhaps nearly impossible—to deal with their trauma. And so, we need to bring compassion rather than an expectation to heal to the forefront.

Second, Walter's story arc is in no way meant to suggest that "racism is a good thing." That would be a gross misinterpretation. The impacts of racism can be horrible, crippling, and murderous, with intergenerational impacts. As anti-Indigenous and anti-Black racism suggests, in North America and beyond we are still living with the legacy of colonization and the transatlantic slave trade.

Yet racism is also a part of our human experience. And to change it, we first have to accept its reality and understand it. The early anti-slavery campaigns started in the late 1700s, which means we've been working toward racial equality, in one form or another, for about 250 years. And our work is not done. Sometimes the more we learn, the more complicated things become. We now know that preferences for our in-groups have been neurologically hardwired into our species—*Homo sapiens*—from the start, approximately 200,000 years ago. So what we're trying to undo, depending on what you consider the starting point, is somewhere between 200 and 200,000 years of history.

We are collectively still traveling along this incredible timeline. And much of the journey to end racism and nurture inclusion is similar for members of dominant and nondominant groups. However, as we saw in the two chapters on power, the day-to-day realities and challenges faced by dominant and nondominant people can be very different, whether we are aware of these differences or not. Therefore, it follows that the paths toward healing the racial rifts in society will also have to be somewhat different. Walter's story is one of a specific journey faced by a member of a nondominant group. The following story of Francine will help us view things from a dominant group's perspective.

The First Step to Reconciliation

In the summer of 1990, a blockade was set up on a local road to Kanesatake, home to the mostly English-speaking Mohawk people in the French-speaking province of Quebec, about an hour's drive from Montreal.[7] Once again, Indigenous rights were being trampled. This time the scenario involved the local town, Oka, pushing for townhouse development and expansion plans for a golf course over ancestral burial grounds without consent or permission from the Mohawks of the Kanesatake community.

After spending a frustrating year going through official legal channels and protesting peacefully, the Mohawk community mobilized to protect their land. An armed standoff ensued with the Quebec provincial police, and eventually the army, which lasted seventy-eight days. It was dramatic and tense, and a skirmish of gunfire occurred during a botched police raid on a Mohawk barricade. One police officer, Marcel Lemay, was shot and killed. Bereaved were his pregnant wife, two-year-old daughter, and sister, Francine, along with the rest of the Lemay family. Photographs capturing solemn-faced Francine Lemay walking behind her thirty-one-year-old brother's casket were all over the news.

Fourteen years after the crisis, two university students asked Francine for her opinion on this historical event.[8] Embarrassed to have no opinion to offer, she read an account of the history of Oka. Curiously, the book that made it into her hands—*At the Woods' Edge*—was powerful but not well known, a self-published historical overview written by two members of the Mohawk community. It chronicled the First Nations people from Kanesatake's struggle to survive, including the betrayal of treaties and land rights that implicated both French and English governments and the church, which was to have held the land in trust but instead sold it off to settlers.[9]

Francine was shocked to confront this perspective on history.[10] For example, she read of the relocation attempt in 1811 by the federal government. Mohawk families were promised food for the winter and seeds for the spring if they resettled far away in central Ontario. The families that went were only given food for two weeks and lived in tents during the cold, hard winter. Many died of disease and hunger.

This version of historical events stood in stark opposition to what she had learned as a child in Quebec schools. She had been taught about the Iroquois Confederacy—of which the Mohawk Nation was one of six cofounders—whose communities had been portrayed as the "bad Indians" for having supported the British during a pivotal historical period that led to the English victory over the region. The Huron, in contrast, who were Indigenous communities that supported the French, were considered "good Indians." Beyond that, her knowledge of Indigenous history came from Hollywood movies.

"That book changed my life," Lemay told a news reporter. "It really touched my heart, to find out all the injustice, the pain and hurt, all the mistreatment [the Mohawks] received, and the inertia of the government."[11]

As a Christian, she struggled with the contradictions and gaps in her knowledge. As it happened, the same week she learned this new history, a delegation from Kanesatake was at her church to give a presentation about a project they were working on. During the event, Francine found herself shaking all over as she listened. At the end of the presentation, she stood up in front of the congregation and asked to speak.

She shared, "I'm Marcel Lemay's sister, the police officer who got shot at Oka."

Pin drop silence.

She then apologized for the injustices their community had endured, including the racist portrayal by the media.

Mavis Etienne, who had been a key negotiator during the Oka crisis, happened to be part of the Mohawk delegation. Mavis walked over to Francine and hugged her, offering condolences for the loss of her brother.

Could there have been a dry eye in the house?

This was a powerful moment of reconciliation. It would be the first of many steps Francine would take toward healing with the Mohawk people of Kanesatake. Mavis invited Francine to attend an ecumenical gathering to pray for peace and reconciliation, called the Trail of Prayers. Francine went. When the trail ended in the Pines, the site of her brother's death, Francine was overtaken by nausea and weakness. Though she was invited to leave if she needed to, Francine insisted on completing the ceremony.

Francine described this experience as "the first phase of my healing.... I let myself mourn my brother—for the first time."[12] This was in 2004, almost a decade and a half after Marcel's death.

Francine began occasionally attending Mavis Etienne's church in Kanesatake. She also realized that many friends in her francophone community still held old ideas and prejudices. Mohawks were stereotypically referred to as "savages," and as involved in "illegal cigarettes, the bingos, the lottery."[13] She decided she could do something to help educate her fellow Quebecer francophones: translate *At the Woods' Edge* into French. Translation was her profession, and this was her way of giving something back.

The offer was met with surprise, joy, and even some discomfort from the Mohawk community. But they accepted her offer, and she toiled away for nine months until the book was successfully translated. *À l'Orée des Bois* was available in time for the twentieth anniversary of the Oka crisis.

"This is like my contribution for the pain the Mohawks endured throughout the centuries, my way to make amends," she explained. "The first step to reconciliation is knowledge, information. So I have to inform the Québécois, the Francophones about the history of Kanesatake."[14]

A Dominant Group Member's Journey

Francine, being white, needed to go through a process of both learning and unlearning the impacts of racism. It started with her knowledge of historical events. Her teachers had taught her a skewed version of history, one that stereotyped Mohawk people as uncivilized and backward, whose only role was to take a side in the "more important" struggle between two dominant European powers. The assumption was that Indigenous Peoples have no important place in history or contemporary society on their own merit and should be discussed only in reference to one of Canada's "founding" nations. A problematic assumption, to say the least.

This, of course, is not about Francine or her local context. Most non-Indigenous people are taught this version of history through both formal (schools) and informal (news, movies) education sources. The power to write, tell, and reinforce history is a formidable manifestation of institutional power. It was something that Francine took for granted until she started to expand her knowledge of history through another perspective.

Francine modeled the use of key inner skills and took a number of important steps that are relevant for dominant group members.

- **Self-awareness:** Although Francine was prompted by two students who wanted her opinion on the Oka crisis, she had enough self-awareness to realize that she didn't know very much about the topic. That required significant humility.

- **Self-regulation:** Her brother had died. To be asked an opinion about an incident that is so emotionally loaded and not just lash out, point fingers, and verbally attack those she might have held responsible took willpower and fortitude. On the flip side, Francine could also have shut down emotionally, suppressing and avoiding the issue altogether. Managing her emotions and psychological state in a constructive way enabled her to move forward. It also speaks to a level of inner power she already possessed.

- **Self-education and discernment:** Francine could easily have relied on stereotypes and media sound bites to make sense of things during that hot Oka summer. Instead, she had a deep desire to learn more. Her curiosity was coupled with being open to a new perspective on history—one that challenged, and even implicated, her white, European roots. She could easily have taken a defensive posture, dismissed what she was reading, and found resources about Oka that supported her established views and position. But she developed discernment through remaining open and being curious instead of reactive, again a testament to her inner strength.

- **Empathy and relationship-building:** Things could have ended at self-education, but Francine went beyond that. She spoke out publicly, demonstrating empathy for the historical and current struggles of the Mohawk people. In a profound act of reconciliation, she apologized, when it was she who had lost a brother. This triggered another series of events, including condolences from Mohawk representatives. Eventually, she would be able to authentically grieve the loss of her sibling. The ongoing relationship continued with her act of solidarity with the Mohawk people. She translated the book that had educated her into French, as a way of combating stereotypes and educating her own people.

- **Meaning-making:** Marcel Lemay's murderer was never identified nor brought to justice. Francine had to find some way to make sense of her life and of her brother's death. Her strong roots in the Christian faith helped her in some important ways, including her awareness that she needed to forgive. For example, she had gone to Oka shortly after her brother's death to let the Mohawks know she forgave them, but at that time she had not been allowed to cross the police line.[15] Later, Francine acknowledged it had been an inauthentic attempt—it felt more like something she should do as a good Christian. Her first true grieving happened fourteen years later, when she broke down during the Trail of Prayers ceremony. She had to face her mixed emotions, even as she reached out to, and was supported by, the Mohawk community. But her reflections indicate that she found a specific way to make meaning that provides some solace. "You cannot measure someone's pain. I cannot say my pain is greater than what the Mohawks went through, through centuries of abuse," she said during a radio interview, her voice sounding reflective. "But through his death, I found friends."[16]

A Nondominant Group Member's Journey

Walter's story is intimately tied to the journey of Grassy Narrows. I don't want to speculate on his personal struggles and journey, as I never learned those details. But at the least, he was a witness to the decline in his community's choices and possibilities due to various levels of government intervention (or lack thereof) and the rise of destructive elements such as mercury poisoning, unemployment, and addictions. His life was definitely touched in some way, directly or indirectly, by living in a small community that survived such difficult challenges as forced relocation, Minamata disease, and destruction of traditional ways of life. Many direct and indirect influences from outside Grassy Narrows can be seen as sources of the negative events and changes in the community's history.

Yet Walter did not hate whites or non-Indigenous people. Instead he spoke of reconciliation, healing, and relationship. Although the qualities he possessed are similar to Francine's, they come from a different perspective, that of a nondominant group member. We have fewer details of Walter's story, but we can still identify significant aspects of his journey.

- **Meaning-making:** Walter had to make some sense out of the pain that he and his community experienced. Yet when I spoke to him, he wasn't despairing. He spoke with great love, admiration, and respect for Grassy Narrows. He exemplified what it means to be able to see both the beauty and the brutality in the world. He knew the challenges in his community better than anyone. His serenity communicated that he accepted what could not be changed. Equally, by his presence on his community's blockade and pride in their success at resisting logging trucks, he showed a willingness to fight for what could be changed. When speaking with me, he neither sugarcoated the struggles nor drowned in despair. His equanimity and wisdom left me feeling I had met a true elder (although I don't know whether he held that title in his community). A long tradition of healing circles, ceremonies, and other modalities within Indigenous communities can support such healing and the development of inner power.

- **Empathy and relationship-building:** Walter's statement was simple and explicit: "It's not too late for our relationship—there have been mistakes made on both sides." In spite of a lopsided history and an unbalanced dynamic of political and economic power, Walter was willing to take responsibility for mistakes made on his people's side of the relationship. This, too, is a powerful marker of reconciliation. It's a complementary gesture to Francine's apology to the Mohawks, in spite of the loss of her brother.

- **Self-awareness and self-regulation:** Given that Grassy Narrows had been hit with multiple tragedies over many generations, it's not difficult to imagine a person becoming angry, vengeful, violent, or despairing and self-destructive. To be hopeful and whole, Walter had to learn to notice and manage his feelings, to grieve, to express, and to heal. This required him to gain insight into his own behavior and choices and to be able to self-regulate. As discussed in the chapters on power, an awareness of historical injustices can make anger, self-destruction, and distrust the driving feelings for many people. Yet these same feelings of victimization—feelings that may help mobilize around a cause in the short term—can be a defense mechanism for a deeper level of grief and shame that needs to be named and processed in order to heal. It's clear that Walter had delved into these depths and come out the other side, which required great self-knowledge and personal power.

- **Self-education and discernment:** Walter had a very personal relationship to Grassy Narrows that he was able to put into a broader context. Nondominant groups can develop negative evaluations of themselves—in this case, internalizing an anti-Indigenous bias. But being curious about history and understanding the political and economic forces at play—understanding power dynamics—can help bring compassion for one's situation. In the case of a community like Grassy Narrows, it's not hard to imagine that despair and feelings of inadequacy could set in. When struggles are placed into a broader context, the results can be empowering, allowing greater discernment. Walter had knowledge that allowed him to see both the positives and negatives of

Grassy Narrows. He could also acknowledge mistakes made on the side of his own people, even if the balance was lopsided, a further empowering step to fight an attitude of victimization.

Both Walter's and Francine's stories hold many powerful lessons on moving forward from horrifying loss and traumatic racial conflict in an authentic manner. Those lessons go well beyond what has been articulated here. Their experiences may best serve as inspiration rather than recipe. There are no clean lines or steps for recovery or meaning-making. It's a deeply individual process that is messy, painful, and uncertain. Perhaps just knowing that we have to reach both outward to others and inward into ourselves may serve as a starting point.

Intergroup tensions and issues of racial difference are never-ending processes of learning, unlearning, and transformation. A willingness to engage in this process can be painful, but it can also bear fruit in unique ways. It may, at times, permanently shift our attitudes and worldviews, making things better for others and ourselves.

Learning New Implicit Habits

My partner, Annahid, and I were processing our marriage license at one side of the entrance to city hall. Given the four cultural and three religious traditions of our families, we had decided to do a multistep process of "tying the knot," inviting the Muslim, Christian, Hindu, Persian, British, Indian, and Pakistani aspects of our ancestry into a self-designed ceremony. She and I had discussed for many months how to navigate all the competing needs—including our own. We were doing a decent job of juggling the essentials, in spite of dropping a few balls. We had just completed an official civil service on an upper floor of city hall, the first of four ceremonies to take place over a monthlong extravaganza of related events.

The administrator, a white woman who looked fifty-something and spoke with a Scottish-sounding accent, was helping us get through a legal hurdle: the paperwork. Upon completion of the appropriate forms, she turned to Annahid, smiled warmly, and shook her hand.

"Congratulations," she said emphatically. "You're now officially married!" Without missing a beat, the woman turned to me and asked in the most genuine, easeful manner, "Do you shake hands?"

"Uh... yes," I replied, quite surprised at the question.

"Well, congratulations!" she said and shook my hand with equal warmth and kindness.

I never got the woman's name, but that experience has stayed with me. Not because the signing of our marriage papers was memorable—it's a simple bureaucratic procedure, after all—but because of what this woman taught me about diversity and inclusion. Let me explain.

I had completed the paperwork first. After Annahid signed to "seal the deal," the administrator immediately congratulated her with a handshake.

As a bureaucrat who deals with the careful registration of names, she likely did some quick cultural calculus—first, guessing that my name *might* be Muslim in origin, and second, knowing that *some* Muslim men don't shake hands with women. Instead of feeling thrown off by this knowledge, she confronted the uncertainty head-on by asking me a straightforward question: *Do you shake hands?*

Although the question surprised me, it was asked with the utmost ease and comfort. It was clear by the administrator's body language, tone, and mannerisms that she would have been completely comfortable had my response been yes or no.

She demonstrated what is sometimes referred to as the Platinum Rule in the world of inclusion, diversity, and equity: *Treat others the way* they *want to be treated.* (Rather than, as the Golden Rule many of us have learned puts it, the way you want to be treated.)

As discussed previously, shaking hands is such a cultural habit that we may not think twice about it. For many people in mainstream North America—myself included—being uncertain about shaking hands with another person can create feelings of discomfort and anxiety, both consciously and unconsciously. We prefer to avoid feelings associated with making mistakes or doing things wrong. Our shame buttons can get pressed, which usually means anger and resentment are not far behind.

This administrator's actions and attitude demonstrated what it means to expand one's pattern recognition repertoire in order to learn a new cultural habit. And, given her age, she also exemplified that learning is a lifelong process. Many of us old folks can, in fact, learn new tricks.

The key is internal motivation. Somewhere along the way, she had decided it was important to unlearn a long-ingrained habit (assuming that shaking hands was "normal" for everyone) and learn a new practice (asking the question, if she suspected it might be an issue). She was specifically welcoming conservative Muslims in this case, although there are other cultural groups who share the same practice. She had made a choice to become conscious of something that most people take for granted.

That subtle but extremely important step is what the bulk of the Deep Diversity framework is about: changing implicit habits. To nurture inclusion, diversity, and equity, we have to become aware of unconscious behavioral patterns. We have to become more aware of circumstances where such default tendencies do not serve our relationship with others, especially those we perceive to be different than ourselves.

A Breakdown of Skills

No matter how we slice it, the outcome is the same. The less aware we are of unconscious impulses and dynamics, the more impact they have in our lives. Expanding our pattern recognition tool kit using the four lenses of Deep Diversity—emotions, bias, identity, and power—is a good starting point.

Breaking down the story of the marriage license administrator may help make the process more transparent:

- **Identity:** First, she recognized that to do her job well, she needed to tune in to the needs of members of other groups, especially those who are historically nondominant, some of whom may have different habits than she was accustomed to.

- **Emotions:** She managed her emotions and opted to be curious rather than anxious, angry, or resentful about the need to accommodate, challenging her own assumptions about handshakes with members of the opposite sex. She also likely had to go through a series of interactions where she was less successful and more awkward—this is true anytime we learn something new—and, again, dealt with the feelings related to failure. Rather than giving up, denouncing, or demanding that "they" conform to "our" standards, she stood firm in her commitment to learning and was resilient enough to develop a new way of doing things.

- **Bias:** In taking a new approach, she also had to tackle the preference or bias of her in-group, which would have predisposed her to think about a "right" versus "wrong" way of congratulating someone.

- **Power:** Finally, there was an implicit and explicit use of power to create positive outcomes for everyone involved, including herself. She mobilized positional power within the institution to help create a more welcoming interaction and experience for nondominant groups. With or without intention, she used her dominant group identity as a white person to be an ally to a nondominant group member, leveraging her higher racial rank to create an inclusive interaction. The inner work she did to create a new way of doing things—including being curious, managing her emotions, going through a learning curve, and making mistakes—required the use of her personal power. The results came across in the ease with which she interacted with me regarding the handshake. The outcomes also benefit her. She was more comfortable and could tilt toward (rather than away from) interactions with out-group members. Further, her fluidity in dealing with newness coincidentally notched up her personal power.

We're talking about changing our habits at the implicit level. It's very subtle but extremely important work. Collectively, our implicit biases and attitudes favor the dominant in-groups—white people when talking about race and men when talking about gender—reinforcing social inequities. From seemingly innocent preferences such as which stranger we choose to sit beside on public transit, to critical issues in health care and policing that can result in people losing their lives, our individual implicit biases and related unconscious institutional habits play a significant role.

The Deep Diversity lens can help us name, talk about, and navigate systemic discrimination. Asking the four gateway questions can help us uncover issues that may be more hidden or unconscious:

- What are the influences of emotions in this situation, group, or issue?
- What are the influences of bias?
- What are the influences of tribes?
- What are the influences of power?

The related skills—self-awareness, self-regulation, empathy, discernment, relationship management, and conflict competence—are the critical inner-work tools we need to sharpen for this journey. Meditation is the practice that can deepen our inner skills, and compassion is the hidden fuel needed to enhance our ability to make and accept the mistakes that necessarily have to be made as we learn about others.

A Final Word

In a post-Trump context, we are at a critical moment in history not just for the U.S. but for all democratic Western nations. As mentioned previously, Trump triggered a "global racist contagion" with hate crimes, xenophobia, and nationalism skyrocketing between 2016 and 2020. Trump is not the cause, but rather the most cancerous symptom of a problem that has been decades in the making. We are living through a time of authoritarian-minded leaders in some of the most powerful and important geopolitical nations—Xi Jinping (China), Vladimir Putin (Russia), Narendra Modi (India), Jair Bolsonaro (Brazil), Recep Erdoğan (Turkey), Rodrigo Duterte (Philippines).

And, unfortunately, the conditions for authoritarianism have not dissipated in the U.S. with the Joe Biden presidency. At the time of finalizing this book, most average Republican voters still believed the "Big Lie" that the 2020 election was "stolen" from Trump, and that he was not responsible for the insurrectionist mob that attacked the Capitol Building on January 6, 2021. The vast majority of elected Republican leaders not only refused to sanction him in any way but instead went the other way: threatening and ousting any party members who dared criticize him publicly. Trump's influence on them has only increased since he officially left office, as evidenced by Republicans shutting down the creation of a bipartisan commission to investigate the January 6 attacks in an effort to suffocate the full truth from emerging, and, more ominously, suggesting that far-right extremism has firmly taken over the party of Lincoln.[17]

My strong belief is that we cannot eliminate racism and oppression if we do not live in a democracy. And U.S. democracy itself is under attack in a way that we've not experienced since 1860, just before the American Civil War. What researchers of authoritarianism have found is that the death of democracies occurs when two sides are so polarized that people see each other as enemies.[18] We are at a critical juncture, what john powell from the Othering & Belonging Institute describes as the moment of *bridging* or *breaking*.[19] We either bridge toward a more just, unified world or we fragment into civil conflict and cycles of recrimination. The former is the path of healing and hope; the latter is the path of mutual self-destruction and despair.

So the time for bridge-builders is now!

We need all hands on deck to reduce the divide. There are millions of jobs to do. Find the one that calls you or that you are good at—everything is needed. If you can organize people or protests, then do so. If you can be an advocate for yourself or others, do so. If you are a researcher, then do research. If you are called to look after your people, then tend the home fires. If you are an educator, then teach!

Change is always possible, even against the greatest odds. Sometimes all you need is the strength to offer dinner and dialogue.

I leave you with a closing story about two people: Derek Black and Matthew Stevenson.[20] Derek Black grew up in a white nationalist family, and is the godson of David Duke, former KKK grand wizard. Derek never went to public school but was instead literally homeschooled in hate. His father founded the white extremist website Stormfront, and by age eleven, Derek had already created a kids' page and was touting the white extremist message. By the time he was twenty, Derek was seen as the heir to the white nationalist throne.

Derek wanted to study history—to be a smart white nationalist, to understand the purity of the white race—so he enrolled in a liberal arts college in Florida. He learned really quickly that diversity was valued and hate not tolerated, so he stayed silent on his views and tried to blend in. He lived a double life, hanging out with college acquaintances while also calling into this dad's weekly radio show spouting the ugliness of white supremacy. After a couple of semesters, as with all deceptions, Derek's secret was finally revealed and he became shunned on campus, a pariah.

Enter into the picture Matthew Stevenson, an Orthodox Jewish student at the same college who traveled the same circles as Derek on campus. As one of the few Orthodox Jewish students on campus, he organized and cooked a weekly Shabbat dinner at his apartment, inviting friends both Jewish and non-Jewish. When Derek's cover was blown on campus, Matthew's response was, "What happens if it's not too late? Maybe there's still hope for Derek."

And with that, Matthew, an Orthodox Jew, invited Derek, the young king-to-be of white nationalism, to Shabbat dinner.

And Derek, because he was socially isolated, accepted. Shabbat dinner that night was only a small group instead of the usual ten to twelve people and, as you might imagine, uncomfortable. Under strict instruction from Matthew, white nationalism was not discussed. Instead, they talked about religion, an area of common interest. And so, Derek was invited back the following week. And the week after.

Slowly, over time, the conversations began to get into what Derek believed and why. He showed up with his research, and Matthew and his friends with theirs. They debated the issues, and slowly over time, Derek's beliefs started to change. First a little, then to such a degree that after two years of Shabbat dinners, he renounced his beliefs, publicly alienating himself from his family and childhood. And today, Derek says his biggest desire is to disappear from public life and never be heard from again. But he can't—because he knows he's done too much harm and that reparations need to be made.

To me, this example, though extreme, is helpful and instructive. Matthew Stevenson embodies a way of being that comes of being rooted in deep personal power. If change is possible for someone like Derek—whose belief system was primarily built on anti-Semitism and racial hatred—is it too much for us to think that building bridges with colleagues, neighbors, or family members with whom we politically disagree might be worth the effort?

RACISM REMAINS a defining issue in our world, and tackling it on a systemic level is more subtle and tricky than confronting its more overt manifestations. But we all need to make meaning of something that is inherently awful, unfair, and at times violent.

I have grappled with the themes of prejudice and discrimination for twenty-five years now, as a teacher, consultant, and lifelong student. And I continue to struggle with them. There is no perfect answer to racism, but this book outlines my thinking so far. (And I'm sure I've raised more questions than I've answered.)

When I'm really in a difficult emotional place because something terrible has happened—like the insurrectionist mob violence of January 6 or the murder of East Asian women in Atlanta—I try to put this work into a historical context.

We are here because of all those who came before us—because of the trials, tribulations, mistakes, and successes of our ancestors. The civil rights movement tore through the poisonous outer bark of racism. As a result, overt forms of discrimination are rejected by the mainstream today.

Enough has changed that during my adult life, a Black man with humble beginnings became the president of the United States and a Black Asian woman the vice president. But our work is not complete. Perhaps, like democracy, it is a project with no real end. Perhaps it is only a series of ongoing steps we take, with improvements and renovations outweighing the setbacks. We inherited this responsibility from our elders, and we must continue moving it forward.

I also draw on my belief that we, as a species, are learners. We have changed our thinking and behaviors over generations and will continue to do so. We are far from where we need to be, but certainly further ahead than when I was born in the late 1960s, despite the different risks we face today.

When I am overwhelmed by events or frustrated that change is not happening quickly enough, I make meaning by trying to zoom out and take a bird's-eye view of this shared journey we are on. From this vantage point, we're advancing. I also recall that the ratio of blood, sweat, and tears has shifted. Less blood is required than in our ancestors' time. But an unending supply of sweat and tears is needed to continue this audacious project of creating the world we want, one where fairness, reconciliation, and repair are more than a figment of our imaginations.

As many are beginning to recognize, democracy is less a noun than a verb—it is an ongoing practice for a nation rather than a product that is possessed automatically because people get to vote. Translated from Greek as "rule by the people," democracy is just straight-up hard, messy work of bringing the voices from the margins of society to the center, of preventing domination of minority groups by the majority. As we are currently witnessing, if democracy is taken for granted and left untended, it can be threatened at any point, even in the richest countries with the longest democratic traditions.

Tackling our unconscious biases individually and our systemic racial habits collectively is part of this ongoing historical project, of what Dr. Martin Luther King Jr. famously referred to as the Beloved Community, where our children and the generations to come have the tools, resources, and opportunities to be their best, most luminous selves. Compassion is the driving force for me—as well as understanding that I am a small part of a journey that started long before I got here and will continue long after I am gone. This is not a sprint, but an ultramarathon, spanning generations.

At least, that is how I make constructive sense of my involvement in the work on race, diversity, and our differences.

If you haven't already done so, you're invited to begin your personal meaning-making project, whether you're Black, Indigenous, a person of color, or white. My deepest desire is that in some small way, this book will help you constructively deal with our shared challenges.

Salaam. Namaste.

Acknowledgments

ANYTHING WE WRITE or create is really a snapshot in time of who we are and the ever-changing circumstances that surround us. Such is the case with me on the cusp of the revised edition of *Deep Diversity*, six years after its original publication in 2015.

So what's changed since?

The world has, for one. Trump. COVID-19. George Floyd.

I have, for another.

I'm unsure what the experiences of other authors are, but I feel it took me six years to "grow into" the book I originally wrote. Looking back, I was nervous and a bit timid about what I wanted to say. Although the Deep Diversity framework had been the result of road testing and refining for almost two decades at that point, it was still a very unconventional approach to addressing oppression, especially among my social justice peers.

But taking these ideas public around the U.S. and Canada since then has added a lot more experience points, providing considerable time to learn, to grow, and to be challenged as well as affirmed.

Following the George Floyd protests, the concept of systemic racism went mainstream. So it's time to talk about how to teach that idea well. Furthermore, Donald Trump dropped us at the extreme edge of democracy-destroying polarization, and those of us with civil war histories in our backgrounds recognize the precarity of this era, that mutual self-destruction lies at the bottom of that cliff. It's therefore critical to talk about compassion and how to do racial justice work with polarity awareness, with emotional and psychological fluidity, rather than rigidity and binary thinking.

So I'd like to start by naming and thanking those people who have helped me grow into the new version of my book.

This begins by acknowledging my stellar agents, Samantha Haywood and Léonicka Valcius, the dynamic duo who have worked tirelessly to bring this book to a larger audience. Special thanks to you, Sam, for your warmth and belief in my work, for inviting me to join Transatlantic Agency after our coffee shop meeting back in December 2018.

All writers need a good publisher, and I'm lucky to have found the fabulous team at Greystone Books. Special appreciation to Jen Croll, my editor, who recognized the need to publish this revised edition; Megan Jones, who's done a great job spearheading publicity; and Jess Sullivan for the many cover designs! As well, I'd like to thank Jess Schulman, Meg Yamamoto, and Alison Strobel for copyediting, proofreading, and indexing. Thanks also to my first publisher, Between the Lines, for helping get the original edition of *Deep Diversity* out into the world.

Heartfelt thanks to Barb Thomas and D'arcy Martin, who consistently show up with such love and care, from pampering Annahid and me with exquisite backyard dinners, to engaging our spirited kids, to attending our trainings, to being deep coconspirators.

Incredible gratitude to Medria Connolly and Bryan Nichols, who supported me and traveled with me around the U.S. conference circuit. Who knew that our chance meeting in L.A. would turn into such a wonderful collaboration and celebratory friendship?

To the dear friends who were part of the Grief Circle that helped Annahid and me make it through the pandemic emotionally: Nick, Vanessa, Sandy, Parker, Karen, Annie, Nate, Barb, and D'arcy. Our gatherings have helped me understand what Beloved Community could look like in real time.

To all the wonderful friends and colleagues in Madison, Wisconsin, my home base in the U.S.; specifically, to Geraldine Paredes

Vásquez, Lisa Baker, and Steve Gilchrist, who have been incredibly supportive, helping spread Deep Diversity in that context. And to the fabulous team at University of Wisconsin Credit Union, including Pam Peterson, Sheila Milton, Rob Van Nevel, and Paul Kundert—your leadership has been inspirational and I've learned so much through our collaboration over the last five years.

Thanks also to the researchers who provided support and feedback, including Mahzarin Banaji at Harvard University, who took the time out of her outlandish schedule not only to meet with me but also to provide detailed notes on my original manuscript; Kerry Kawakami at York University; and Michael Inzlicht at the University of Toronto Scarborough, as well as author/psychologist Rick Hanson, founder of the Wellspring Institute for Neuroscience and Contemplative Wisdom.

To Julie Diamond for her incredible friendship and groundbreaking work on power; Rhonda Magee for inviting us into the inner work needed for racial justice; and Loretta Ross for her outrageous courage and trailblazing justice efforts, including the challenge to nurture cultures that call in rather than call out . . . as well as general badassery.

To all the smart white, privileged men in high places: interacting with you made me realize that, even with the best of intentions, there is so much you can't "see" because of internalized dominance, and helped me birth the idea of *racial pattern recognition*.

To the warriors of social justice, the important challenges you raised helped me recognize more clearly that the stuck places in activism stem from an overreliance on socioeconomic historical perspectives, helping me name the necessity for *psychological pattern recognition*.

Special thanks to James Beaton, my dear colleague and friend at Anima Leadership, the unsung hero of this writing project. He did what he always does, which is to handle the behind-the-scenes work tirelessly and without complaint, including being researcher extraordinaire and general sounding board. And deep appreciation to other members of the Anima team who have assisted in various ways over the last few years.

Thanks also to my tight-knit family for all the thousands of small and large expressions of support, including childcare, making meals, and celebrating the project milestones. This includes my mom and dad, Saeeda and Anil Choudhury, sisters Bipasha and Monika, brother-in-law Sundeep, and nephews Zephan and Dastan. It's important to also name my extended family, including auntie-cum-second-mom Ummi, my older cousins Saif Bhai and Shams Bhai, and their families, as my Pakistani-Indian heritage is responsible for teaching me the deep art of hosting, something I didn't realize was a signature in my professional work until recently.

I can't leave here without mentioning my partner in both life and business, Annahid Dashtgard. Throughout both iterations of this book she has been the voice of positivity, encouraging me through thick and thin to keep going. Even when we hit tough moments, and I was crabby or graceless, Annahid never questioned the importance of the project or the time it was taking. She is a fearless advocate for the book and its contents, more than I can ever be. Thanks for your generosity and patience, for tending the home fire and lovingly looking after the kids and our business while I was in my writing cave.

Lastly, I want to acknowledge with deepest gratitude the thousands of people I've had the pleasure of speaking to about these world-changing issues, and learning with, over the years. This includes the Anima Leadership community, those incredible people who keep coming back to our trainings, conferences, and cafés. Through each interaction, whether live or virtual, I leave inspired by the many individuals working tirelessly toward the same goals of justice, equity, and democracy. May this book be another step on our collective path.

Notes

Preface

1 Coronavirus Research Center, Johns Hopkins University & Medicine, accessed November 30, 2020, https://coronavirus.jhu.edu/map.html.

2 Mary Jo Murphy, "Thirteen Writers from Around the World on the Way the President Scrambled Their Nations' Affairs," *Washington Post*, December 19, 2019.

3 Rohan Khazanchi, Charlesnika T. Evans, and Jasmine R. Marcelin, "Racism, Not Race, Drives Inequity across the COVID-19 Continuum," JAMA Network Open, September 25, 2020; Peel Health Surveillance, *COVID-19 and Social Determinants of Health: Race and Occupation*, Region of Peel (Ontario), August 7, 2020.

4 Steven Levitsky and Daniel Ziblatt, *How Democracies Die* (New York: Crown Publishing, 2018).

5 Othering & Belonging Institute, "Bridging —Towards A Society Built on Belonging: Animated Video + Curriculum," November 14, 2018, https://belonging.berkeley.edu/bridging-towards-society-built-belonging-animated-video-curriculum.

6 Four basic approaches to supporting racial, ethnic, and cultural cohesion:

1. Multiculturalism: The sharing of cultural foods and celebrations, including dance, songs, and art. This approach is easy to do, engaging, and fun but pretty laissez-faire about making concrete social change.
2. Cross-cultural communication/cultural competence: Learning about cultural customs, norms, and differences in communications styles to achieve specific goals. For example, Mexicans favor relationship-building before getting down to business deals, Asian people avoid eye contact as a sign of respect, or Muslims don't shake hands with members of the opposite sex. These strategies can help build an understanding of certain customs within ethnocultural groups but frequently rely on generalizations

that are a bit too fixed, reinforcing stereotypes.

3. Business case for diversity: In the context of organizations, this approach focuses on the bottom-line benefits of a diverse, inclusive workforce. This strategy is beneficial in creating buy-in by senior decision makers but tends to candy-coat issues of discrimination and racism, usually avoiding the difficult issues.
4. Anti-racism/anti-oppression (ARAO): From this perspective, discrimination and racism are systemic; therefore understanding the dynamics of power and privilege between white and nonwhite peoples—Black, Indigenous, people of color—is essential to creating racial inclusion. This approach has a good ability to predict unconscious behavior between dominant and non-dominant groups and identifies the redistribution of power as key to making change. The theory suggests that to be nonracist within an unjust system only maintains the status quo; therefore we must seek to be *anti*-racist in our words, actions, and thinking. *Anti-oppression* is the umbrella term that includes other key identity factors related to systemic forms of discrimination such as sexism, heterosexism, ableism, and classism. Traditional approaches to ARAO, however, often shut people down with hard-edge strategies that over-rely on "shame and blame" to create change.

ONE: The Four Pillars of Deep Diversity

1 Some names have been changed to protect the identity of those involved.

2 The racial slur used by the white student that specifically targets Black people is today considered so offensive that it shouldn't even appear in print.

3 K. Kawakami, E. Dunn, F. Karmali, and J.F. Dovidio, "Mispredicting Affective and Behavioral Responses to Racism," *Science* 323 (2009): 276–78.

4 K. Kawakami, interview with Shakil Choudhury, October 18, 2012, York University Department of Psychology, Toronto.

5 Francine Karmali, Kerry Kawakami, and Elizabeth Page-Gould, "He Said What? Physiological and Cognitive Responses to Imagining and Witnessing Outgroup Racism," *Journal of Experimental Psychology: General* 146, no. 8 (2017): 1073–85.

6 Thomas Lewis, Fari Amini, and Richard Lannon, *A General Theory of Love* (New York: Vintage Books, 2000), 35–65.

7 Lewis, Amini, and Lannon, 36.

8 Matthew D. Lieberman, *Social: Why Our Brains Are Wired to Connect* (New York: Crown Publishers, 2013), 39–54.

9 Mahzarin R. Banaji and Anthony G. Greenwald, *Blind Spot: Hidden Biases of Good People* (New York: Delacorte Press, 2013), 32–52.

10 M.R. Banaji and R. Bhaskar, "Implicit Stereotypes and Memory: The Bounded Rationality of Social Beliefs," in *Memory, Brain, and Belief*,

eds. D.L. Schacter and E. Scarry (Cambridge, MA: Harvard University Press, 2000), 139–75.

11 Brian A. Nosek, Mahzarin R. Banaji, and Anthony G. Greenwald, "Harvesting Implicit Group Attitudes and Beliefs from a Demonstration Web Site," *Group Dynamics: Theory, Research, and Practice* 6, no. 1 (2002): 101–15.

12 Jennifer N. Gutsell and Michael Inzlicht, "Empathy Constrained: Prejudice Predicts Reduced Mental Simulation of Actions during Observation of Outgroups," *Journal of Experimental Social Psychology* 46 (2010): 841–45.

13 ScienceDaily, "Human Brain Recognizes and Reacts to Race," April 7, 2010, www.sciencedaily.com.

14 Shihui Han, "Neurocognitive Basis of Racial Ingroup Bias in Empathy," *Trends in Cognitive Sciences*, May 1, 2018; Bradley D. Mattan, Kevin Y. Wei, Jasmin Cloutier, and Jennifer T. Kubota, "The Social Neuroscience of Race-Based and Status-Based Prejudice," *Current Opinion in Psychology*, December 2018.

15 John Bowden, "Cuccinelli Says George Floyd's Death Wasn't about Race," *The Hill*, June 8, 2020.

16 Tom Blackwell, "Ferguson Mayor Says He Was Unaware of Racial 'Frustrations' in Community until Michael Brown Shooting," *National Post*, November 26, 2014.

17 Konrad Yakabuski, "Trayvon's Killing Echoes an Uglier Time in America," *Globe and Mail* (Toronto), March 30, 2013.

18 Frank Newport, "Blacks, Nonblacks Hold Sharply Different Views of Martin Case: Blacks More Likely to Believe Race Is a Major Factor," *Gallup Politics*, April 5, 2012.

19 ABC News, "74% of Americans View George Floyd's Death as an Underlying Racial Injustice Problem: Poll," June 5, 2020.

20 Jeffrey M. Jones, "Black, White Adults' Confidence Diverges Most on Police," *Gallup*, August 12, 2020.

21 Ashley Southall, "Commissioner Denies Racial Bias in Social Distancing Policing," *New York Times*, May 12, 2020.

22 BBC News, "Ahmaud Arbery: Father and Son Charged with Murder of U.S. Black Jogger," May 8, 2020.

23 Eric Levenson, "A Timeline of Breonna Taylor's Case Since Police Broke Down Her Door and Shot Her," *CNN*, September 24, 2020.

24 Chris Stewart, "Metis Hunter in Alberta Says Threats Not New in Province," *APTN National News*, April 6, 2020.

25 "Remembering Joyce Echaquan: Frequently Asked Questions and the Facts So Far," *APTN National News*, October 7, 2020.

26 Maureen J. Brown, *We Are Not Alone: Police Racial Profiling in Canada, the United States, and the United Kingdom* (Toronto: African Canadian Community Coalition on Racial Profiling, 2007), 9.

27 Connie Hasset-Walker, "The Racist Roots of American Policing: From Slave Patrols to Traffic Stops," *The Conversation*, June 4, 2019.

28 Jonathan Haidt, *The Happiness Hypothesis: Finding Modern Truth in Ancient Wisdom* (New York: Basic Books, 2006), 43.

29 Feng Sheng and Shihui Han, "Manipulations of Cognitive Strategies and Intergroup Relationships Reduce the Racial Bias in Empathic Neural Responses," *NeuroImage* 61 (2012): 786–97.
30 Dora Capozza, Luca Andrighetto, Gian Antonio Di Bernardo, and Rosella Falvo, "Does Status Affect Intergroup Perceptions of Humanity?," *Group Processes & Intergroup Relations* 15, no. 3 (2012): 363–77.
31 David Dobbs, "Mastery of Emotions," *Scientific American Mind* (February/March 2006): 48.
32 Dobbs, 44–49.
33 Daniel J. Siegel, *Mindsight: The New Science of Personal Transformation* (New York: Bantam Books Trade Paperbacks, 2011), 133.
34 Siegel, 198–200.
35 Harold Garfinkel, *Studies in Ethnomethodology* (Cambridge, U.K.: Polity Press, 1991), 36.
36 Banaji and Bhaskar, "Implicit Stereotypes and Memory," 142–43.
37 Siegel, *Mindsight*, 24–25.
38 Haidt, *The Happiness Hypothesis*, 37–38.
39 Beck Institute for Cognitive Behavior Therapy, "Cognitive Therapy Can Treat," www.beckinstitute.org.
40 Rick Hanson, *Buddha's Brain: The Practical Neuroscience of Happiness, Love and Wisdom* (Oakland, CA: New Harbinger Publications, 2009), 2–3.
41 Lorne Ladner, *The Lost Art of Compassion: Discovering the Practice of Happiness in the Meeting of Buddhism and Psychology* (San Francisco: HarperCollins, 2004), 14.
42 Edith Eisler, "Yo-Yo Ma: On the Silk Road," *All Things Strings* (May/June 2001), www.allthingsstrings.com.

TWO: Emotions: Understanding Ourselves and Others

1 Town of Hérouxville, http://welcome-to-herouxville-quebec-canada.blogspot.com.
2 Lucas Powers, "Conservatives Pledge Funds, Tip Line to Combat 'Barbaric Cultural Practices,'" *CBC News*, October 2, 2015.
3 Canadian Civil Liberties Association, "Bill 21: The Law against Religious Freedom," accessed December 6, 2020, https://ccla.org/bill-21/.
4 Benjamin Shingler, "Judge Suspends Quebec Face-Covering Ban, Says It Appears to Violate Charter," *CBC News*, June 28, 2018.
5 Marion Scott, "Islamophobia Surging in Quebec since Charter, Group Says: 117 Complaints of Verbal, Physical Abuse Made between Sept. 15 and Oct. 15 Compared with 25 Total for Previous Nine Months," *The Gazette*, November 6, 2013.
6 Tanvi Misra, "The Year in Anti-Immigrant Rhetoric," *Bloomberg CityLab*, December 29, 2015.
7 Jaweed Kaleem, "Latinos and Transgender People See Big Increases in Hate Crimes, FBI Reports," *Los Angeles Times*, November 12, 2019.
8 Marco Giani and Pierre-Guillaume Méon, "Global Racist Contagion Following Donald Trump's Election," Cambridge University Press, December 9, 2019.
9 Martin Armstrong, "Hate Crimes Rising Sharply in England and Wales," *Statista*, October 15, 2019.

10 Lewis, Amini, and Lannon, *A General Theory of Love*, 3.

11 Daniel Goleman, "What Makes a Leader?," *Harvard Business Review* (January 2004), https://hbr.org/2004/01/what-makes-a-leader.

12 Daniel Goleman, Richard E. Boyatzis, and Annie McKee, *Primal Leadership: Learning to Lead with Emotional Intelligence* (Boston: Harvard Business Press, 2002, 2004), 28.

13 Haidt, *The Happiness Hypothesis*, 12.

14 Paul Rozin and Edward B. Royzman, "Negativity Bias, Negativity Dominance, and Contagion," *Personality and Social Psychology Review* 5, no. 4 (2001): 296–320.

15 Hanson, *Buddha's Brain*, 40–42.

16 Lewis, Amini, and Lannon, *A General Theory of Love*, 41.

17 Eunice Yang, David H. Zald, and Randolph Blake, "Fearful Expressions Gain Preferential Access to Awareness during Continuous Flash Suppression," *Emotion* 7, no. 4 (November 2007): 882–86.

18 Jennifer Eberhardt, "Imaging Race," *American Psychologist* 60, no. 2 (February/March 2005): 181–90.

19 City of Brampton Economic Development Office, "National Household Survey Bulletin #1: Immigration, Citizenship, Place of Birth, Language, Ethnic Origin, Visible Minorities, Religion, and Aboriginal Peoples," May 2013, www.brampton.ca.

20 Mark Easton, "Why Have the White British Left London?," *BBC News*, February 13, 2013.

21 San Grewal, "Brampton Suffers Identity Crisis as Newcomers Swell City's Population," *Toronto Star*, May 24, 2013.

22 Goleman, Boyatzis, and McKee, *Primal Leadership*, 5–6.

23 Lewis, Amini, and Lannon, *A General Theory of Love*, 64–65.

24 Sigal Barsade, "Faster Than a Speeding Text: 'Emotional Contagion' at Work," *Psychology Today*, October 15, 2014, https://www.psychologytoday.com/ca/blog/the-science-work/201410/faster-speeding-text-emotional-contagion-work.

25 Lewis, Amini, and Lannon, *A General Theory of Love*, 78–96.

26 Lieberman, *Social*, 40.

27 Lieberman, 57–59.

28 Kipling D. Williams, "The Pain of Exclusion," *Scientific American Mind* (January/February 2011): 30–37.

29 Goleman, Boyatzis, and McKee, *Primal Leadership*, 14.

30 Goleman, Boyatzis, and McKee, 12–14.

31 Cara Williams, "Sources of Workplace Stress," *Perspectives on Labour and Income: The Online Edition* (Statistics Canada) 4, no. 6 (June 2003).

32 Chris Park, "Once Burned Out, Twice Shy: The Unaffordable Cost of Work-Related Stress," *Open Access Government*, August 13, 2019.

33 Goleman, Boyatzis, and McKee, *Primal Leadership*, 5–15.

34 Kevin Dougherty, "Multiculturalism 'Idiocy,' Charges Anti-Immigration Crusader," *Montreal Gazette*, May 20, 2011.

35 Canadian Immigrant Report, "CIReport.ca Interviews: André Drouin," September 16, 2011, www.cireport.ca.

36 David Markowitz, "Trump Is Lying More Than Ever: Just Look at the Data," *Forbes*, May 5, 2020.

37 Levitsky and Ziblatt, *How Democracies Die*.

38 Lewis, Amini, and Lannon, *A General Theory of Love*, 20–33.

39 Lewis, Amini, and Lannon, 20–33.

40 Frank Krueger et al., "The Neural Bases of Key Competencies of Emotional Intelligence," *Proceedings of the National Academy of Sciences of the United States of America* 106, no. 52 (2009): 22486–91.

41 Haidt, *The Happiness Hypothesis*, 29–34.

42 Hanson, *Buddha's Brain*, 96–97.

43 Jaclyn Ronquillo et al., "The Effects of Skin Tone on Race-Related Amygdala Activity: An fMRI Investigation," *Social Cognitive and Affective Neuroscience* 2 (2007): 39–44.

44 Hanson, *Buddha's Brain*, 101.

45 Peter Turchin, *Ages of Discord: A Structural-Demographic Analysis of American History* (Chaplin, CT: Beresta Books, 2016).

46 Siegel, *Mindsight*, xi–xii.

47 Lewis, Amini, and Lannon, *A General Theory of Love*, 41–42.

48 Tiffany A. Ito and Bruce D. Bartholow, "The Neural Correlates of Race," *Trends in Cognitive Sciences* 13, no. 12 (2009): 524–30.

49 Max Weisbuch, Kristin Pauker, and Nalini Ambady, "The Subtle Transmission of Race Bias via Televised Nonverbal Behavior," *Science* 326 (December 2009): 1711–14.

50 Joseph Hall, "TV Clips Reveal Racist Body Language, Study Finds," *Toronto Star*, December 18, 2009, https://www.thestar.com/news/gta/2009/12/18/tv_clips_reveal_racist_body_language_study_finds.html.

51 Michael Inzlicht, interview with Shakil Choudhury, September 17, 2012.

52 Goleman, Boyatzis, and McKee, *Primal Leadership*, 40–45.

53 Goleman, Boyatzis, and McKee, 40.

54 Siegel, *Mindsight*, ix–xii.

55 Amishi P. Jha, "Mindfulness Can Improve Your Attention and Health," *Scientific American Mind* (March/April 2013).

56 Haidt, *The Happiness Hypothesis*; Siegel, *Mindsight*, 60–62.

57 Daniel J. Siegel, *The Mindful Brain: Reflection and Attunement in the Cultivation of Well-Being* (New York: W.W. Norton and Company, 2007), 5–15.

58 Siegel, *Mindsight*, 97–100.

59 Siegel, 86–87.

60 Center for Mindfulness in Medicine, Health Care and Society, University of Massachusetts Medical School, www.umassmed.edu/cfm.

61 Delores B. Lindsey, Richard S. Martinez, and Randall B. Lindsey, *Culturally Proficient Coaching: Supporting Educators to Create Equitable Schools* (Thousand Oaks, CA: Corwin Press, 2007).

THREE: Bias: Prejudice without Awareness

1 Siri Carpenter, "Buried Prejudice," *Scientific American Mind* (April/May 2008): 33–39.

2 Project Implicit, "Frequently Asked Questions: #21 What Is the Difference between 'Implicit' and 'Automatic'?," www.implicit.harvard.edu.

3 D.M. Amodio and S.A. Mendoza, "Implicit Intergroup Bias: Cognitive, Affective, and Motivational Underpinnings," in *Handbook of Implicit Social Cognition*, eds. B. Gawronski and B.K. Payne (New York: Guilford, 2010), 353–74.
4 Mahzarin Banaji, interview with Shakil Choudhury, May 10, 2013, Toronto.
5 Banaji and Bhaskar, "Implicit Stereotypes and Memory," 140.
6 Hanson, *Buddha's Brain*, 6, 31–32.
7 Hanson, 70–73.
8 Timothy Wilson, *Strangers to Ourselves: Discovering the Adaptive Unconscious* (Cambridge, MA: Belknap Press, 2004).
9 Wilson, 24.
10 David G. Myers, "The Powers and Perils of Intuition: Understanding the Nature of Our Gut Instincts," *Scientific American Mind* (June/July 2007): 26.
11 Myers, 25–30.
12 Howard Ross, "Proven Strategies for Addressing Unconscious Bias in the Workplace," *Diversity Best Practices* (2008): 3.
13 Ross.
14 Banaji and Bhaskar, "Implicit Stereotypes and Memory," 140.
15 Curtis D. Hardin and Mahzarin Banaji, "The Nature of Implicit Prejudice: Implications for Personal and Professional Policy," in *The Behavioral Foundations of Public Policy*, ed. Eldar Shafir (Princeton: Princeton University Press, 2012): 13–31.
16 Banaji and Greenwald, *Blind Spot*.
17 Banaji and Bhaskar, "Implicit Stereotypes and Memory," 147.
18 Hardin and Banaji, "The Nature of Implicit Prejudice."
19 Banaji and Bhaskar, "Implicit Stereotypes and Memory," 142–45.
20 Project Implicit, "General Information," www.projectimplicit.net.
21 Capozza et al., "Does Status Affect Intergroup Perceptions of Humanity?," 363–77.
22 Nosek, Banaji, and Greenwald, "Harvesting Implicit Group Attitudes and Beliefs," 101–15.
23 Hardin and Banaji, "The Nature of Implicit Prejudice."
24 Hardin and Banaji.
25 B. Kurdi et al., "Relationship Between the Implicit Association Test and Intergroup Behavior: A Meta-Analysis," *American Psychologist* 74, no. 5 (2019): 569–86.
26 Philip Oreopoulos, "Why Do Skilled Immigrants Struggle in the Labor Market? A Field Experiment with Six Thousand Résumés," University of British Columbia, National Bureau of Economic Research, and Canadian Institute for Advanced Research, June 2009.
27 Shankar Vedantam, "See No Bias," *Washington Post*, January 23, 2005, https://www.washingtonpost.com/archive/lifestyle/magazine/2005/01/23/see-no-bias/a548dee4-4047-4397-a253-f7f780fae575/; also see the original study: Marianne Bertrand and Sendhil Mullainathan, "Are Emily and Greg More Employable Than Lakisha and Jamal? A Field Experiment on Labor Market Discrimination," University of Chicago Graduate School of Business, National

Bureau of Economic Research, July 2003.

28 Lincoln Quillian et al., "Meta-analysis of Field Experiments Shows No Change in Racial Discrimination in Hiring over Time," *Proceedings of the National Academy of Sciences of the United States of America*, September 12, 2017.

29 Narjust Duma et al., "Evaluating Unconscious Bias: Speaker Introductions at an International Oncology Conference," *Journal of Clinical Oncology* 37, no. 15_suppl (May 20, 2019): 10503.

30 Alexander R. Green et al., "Implicit Bias among Physicians and Its Prediction of Thrombolysis Decisions for Black and White Patients," *Journal of General Internal Medicine* 22, no. 9 (2007): 1231–38.

31 Green et al.

32 Tori DeAngelis, "How Does Implicit Bias by Physicians Affect Patients' Health Care?" *Monitor on Psychology* (American Psychological Association) 50, no. 3 (March 2019).

33 Faiza Amin, "Falling Through the Cracks of Canada's Health Care System: The John Rivers Story," *CityNews*, December 4, 2020.

34 Rebecca Hagey et al. (Centre for Equity in Health and Society), *Implementing Accountability for Equity and Ending Racial Backlash in Nursing: Accountability for Systemic Racism Must Be Guaranteed to Uphold Equal Rights in Society and Promote Equity in Health* (Toronto: Canadian Race Relations Foundation, 2005), xxi.

35 Jack Geiger, "Racial Stereotyping and Medicine: The Need for Cultural Competence," *Canadian Medical Association Journal* 164, no. 12 (June 2001): 1699–1700.

36 Carpenter, "Buried Prejudice," 37–39.

37 Hardin and Banaji, "The Nature of Implicit Prejudice."

38 Banaji and Bhaskar, "Implicit Stereotypes and Memory," 147.

39 S. Alexander Haslam et al., "The Social Psychology of Success," *Scientific American Mind* (April/May 2008): 25–26.

40 Haslam et al.

41 Claude M. Steele, *Whistling Vivaldi: How Stereotypes Affect Us and What We Can Do* (New York: W.W. Norton and Company, 2010), 6.

42 Haslam et al.

43 Banaji and Greenwald, *Blind Spot*, 128–30.

44 Ito and Bartholow, "The Neural Correlates of Race," 524–30.

45 Ronquillo et al., "The Effects of Skin Tone on Race-Related Amygdala Activity," 39–44.

46 Margaret Vogel, Alexandra Monesson, and Lisa S. Scott, "Building Biases in Infancy: The Influence of Race on Face and Voice Emotion Matching," *Developmental Science* 15, no. 3 (May 2012): 359–72.

47 David J. Kelly, Paul C. Quinn, Alan M. Slater, Kang Lee, Liezhong Ge, and Olivier Pascalis, "The Other-Race Effect Develops During Infancy: Evidence of Perceptual Narrowing," *Psychological Science* 18, no. 12 (December 2007): 1084–89.

48 Banaji and Greenwald, *Blind Spot*, 128–30.

49 Leda Cosmides and John Tooby, "Evolutionary Psychology: New Perspectives on Cognition and

Motivation," *Annual Review of Psychology* 64 (2013): 201–29.

50 David Pietraszewski, Leda Cosmides, and John Tooby, "The Content of Our Cooperation, Not the Color of Our Skin: An Alliance Detection System Regulates Categorization by Coalition and Race, but Not Sex," *PLOS ONE* 9, no. 2 (2014): e88534.

51 Myers, "The Powers and Perils of Intuition," 25–31.

52 Brandon Stewart, "Bringing Automatic Stereotyping under Control: Implementation Intentions as Efficient Means of Thought Control," *Personality and Social Psychology Bulletin* 34, no. 10 (October 2008): 1334.

53 Norman Doidge, *The Brain That Changes Itself: Stories of Personal Triumph from the Frontiers of Brain Science* (New York: Penguin Books, 2007), 63–64.

54 Carpenter, "Buried Prejudice," 37–38.

55 Carpenter.

56 Inzlicht, interview with Choudhury, September 17, 2012.

57 Carpenter, "Buried Prejudice," 38.

58 Carpenter.

59 J. Kang and M. Banaji, "Fair Measures: A Behavioral Realist Revision of 'Affirmative Action,'" *California Law Review* 94 (2006): 1063–118.

60 Hardin and Banaji, "The Nature of Implicit Prejudice."

61 Markus Brauer, Abdelatif Er-rafiy, Kerry Kawakami, and Curtis E. Phills, "Describing a Group in Positive Terms Reduces Prejudice Less Effectively Than Describing It in Positive and Negative Terms," *Journal of Experimental Social Psychology* 48 (2012): 757–61.

62 Stewart, "Bringing Automatic Stereotyping under Control," 1334.

63 Brauer et al., "Describing a Group in Positive Terms."

64 Feng Sheng and Shihui Han, "Manipulations of Cognitive Strategies and Intergroup Relationships Reduce the Racial Bias in Empathic Neural Responses," *NeuroImage* 61 (2012): 786–97.

65 Lasana T. Harris and Susan T. Fiske, "Social Groups That Elicit Disgust Are Differentially Processed in the mPFC," *Social Cognitive and Affective Neuroscience* 2 (2007): 45–51.

66 Patricia G. Devine et al., "Long-Term Reduction in Implicit Race Bias: A Prejudice Habit-Breaking Intervention," *Journal of Experimental Social Psychology* 48 (2012), 1267–78.

67 Inzlicht, interview with Choudhury, September 17, 2012.

68 Ito and Bartholow, "The Neural Correlates of Race," 524–30.

69 Goleman, Boyatzis, and McKee, *Primal Leadership*, 46.

FOUR: Identity: Belonging Drives Human Behavior

1 Ontario Human Rights Commission (OHRC), "Preliminary Findings: Inquiry into Assaults on Asian Canadian Anglers," December 2007, http://www.ohrc.on.ca/en/preliminary-findings-inquiry-assaults-asian-canadian-anglers.

2 OHRC.

3 OHRC.

4 Myers, "The Powers and Perils of Intuition," 31.

5 American Anthropological Association, "AAA Statement on 'Race,'" May 17, 1998, https://www.americananthro.org/ConnectWithAAA/Content.aspx?ItemNumber=2583.

6 Robert Jensen, *The Heart of Whiteness: Confronting Race, Racism and White Privilege* (San Francisco: City Lights Books, 2005), 14.

7 Human Rights Watch, "Covid-19 Fueling Anti-Asian Racism and Xenophobia Worldwide," May 12, 2020.

8 William B. Gudykunst, *Bridging Differences: Effective Intergroup Communications*, 3rd ed. (Thousand Oaks, CA: Sage Publications, 1998), 40–42.

9 Banaji and Greenwald, *Blind Spot*, 132–33.

10 Gudykunst, *Bridging Differences*, 40–48.

11 Gudykunst, 42.

12 Gudykunst, 14.

13 Jeanne Maglaty, "When Did Girls Start Wearing Pink?," *Smithsonian Magazine*, April 7, 2011, https://www.smithsonianmag.com/arts-culture/when-did-girls-start-wearing-pink-1370097/.

14 Martin Smith, "I Cannot Shake Your Hand, Sir. I'm a Muslim and You're a Man," *Daily Mail Online*, January 20, 2007, https://www.dailymail.co.uk/news/article-430249/I-shake-hand-sir-Im-Muslim-youre-man.html; Andy Levy-Ajzenkopf, "Deputy Mayor Offended by Handshake Snub," *Canadian Jewish News*, February 14, 2008, https://www.cjnews.com/news/deputy-mayor-offended-hand shake-snub; Farooq Sulehria, "And Now the 'Handshake' Issue," *The News International*, March 18, 2010, https://www.thenews.com.pk/archive/print/227503-and-now-the-'hand shake'-issue.

15 Garfinkel, *Studies in Ethnomethodology*, 36–37; Stanley Milgram et al., "Response to Intrusion into Waiting Lines," *Journal of Personality and Social Psychology* 51, no. 4 (1986): 683–89.

16 George Ritzer, *Ethnomethodology in Sociological Theory*, 8th ed. (New York: McGraw-Hill, 2011), 397–98.

17 Tina Lopez and Barb Thomas, *Dancing on Live Embers: Challenging Racism in Organizations* (Toronto: Between the Lines, 2006), 263–72.

18 OHRC, "Preliminary Findings."

19 OHRC.

20 Associated Press, "U.N.: Dozens of Muslims Massacred by Buddhists in Burma," January 24, 2014, https://www.cbsnews.com/news/un-dozens-of-rohingya-muslims-massacred-by-buddhists-in-rakhine-burma/.

21 BBC News Asia, "Pain of Pakistan's Outcast Ahmadis," September 30, 2014, https://www.bbc.com/news/world-asia-29415356.

22 PBS Frontline, "The Triumph of Evil: 100 Days of Slaughter; A Chronology of U.S./U.N. Actions," January 1999, https://www.pbs.org/wgbh/pages/frontline/shows/evil/etc/slaughter.html.

23 *Jimmy Kimmel Live!*, "Clinton Supporters Agree with Donald Trump Quotes," August 4, 2016.

24 Evan Simko-Bednarski, "Court Orders Donald Trump to Pay Legal Fees in Stormy Daniels Suit," *CNN*, August 22, 2020.

25 Maureen Ryan, "The *Access Hollywood* Tape was Vile—and Maybe a Vital Tipping Point," *Vanity Fair*, October 7, 2020.

26 Eliza Relman, "The 26 Who Have Accused Trump of Sexual Misconduct," *Business Insider*, December 2017.

27 Ryan Teague Beckwith, "Here Are All of the Indictments, Guilty Pleas and Convictions from Robert Mueller's Special Investigation," *Time*, March 22, 2019.

28 Aris Folley, "Trump Supporters Whose T-shirts Went Viral: We're Not Traitors," *The Hill*, August 28, 2018.

29 "In Changing U.S. Electorate, Race and Education Remain Stark Dividing Lines," Pew Research Center, June 2, 2020.

30 Thomas E. Mann and Norman J. Ornstein, "Let's Just Say It: The Republicans Are the Problem," *Washington Post*, April 27, 2012; William A. Galston and Thomas E. Mann, "The GOP's Grass-Roots Obstructionists," *Washington Post*, May 16, 2010.

31 Jake Tapper, Lauren Fox, and Veronica Stracqualursi, "Nearly a Dozen Republican Senators Announce Plans to Vote against Counting Electoral Votes," *CNN*, January 2, 2021.

32 Gudykunst, *Bridging Differences*, 31–32.

33 Goleman, Boyatzis, and McKee, *Primal Leadership*, 49–50.

34 C. Daniel Batson and Nadia Y. Ahmad, "Using Empathy to Improve Intergroup Attitudes and Relations," *Social Issues and Policy Review* 3, no. 1 (2009): 141–77.

35 Siegel, *Mindsight*, 60–62.

FIVE: Power: The Dividing Force

1 Jim Sidanius and Felicia Pratto, *Social Dominance: An Intergroup Theory of Hierarchy and Oppression* (New York: Cambridge University Press, 1999).

2 Sidanius and Pratto, 6.

3 Lopez and Thomas, *Dancing on Live Embers*, 269.

4 Sidanius and Pratto, *Social Dominance*, 39–41.

5 Banaji and Greenwald, *Blind Spot*, 140–44.

6 Kirk Makin, "Of 100 New Federally Appointed Judges, 98 Are White, Globe Finds," *Globe and Mail* (Toronto), April 17, 2012.

7 "Statistics regarding Judicial Applicants and Appointees (October 21, 2016 – October 28, 2020)," Office of the Commissioner for Federal Judicial Affairs Canada, Government of Canada.

8 Kirk Makin, "Minority Lawyers Demand Diversity among Appointed Judges," *Globe and Mail* (Toronto), May 8, 2012.

9 Sidanius and Pratto, *Social Dominance*, 41.

10 Sidanius and Pratto, 129.

11 Ta-Nehisi Coates, "The Black Family in the Age of Mass Incarceration," *The Atlantic*, October 2015.

12 John Gramlich, "Black Imprisonment Rate in the US Has Fallen a Third Since 2006," Pew Research Center, May 6, 2020.

13 Jennifer E. Cobbina, "The Links between Slavery, Policing and Racism," *From the Square* (blog), NYU Press, July 30, 2019.

14 "Cleveland Police Handcuff Tamir Rice's Sister after Shooting 12-Year-Old—Video," *Guardian*, January 8, 2015, https://www.theguardian.com/us-news/video/2015/jan/08/

new-video-tamir-rice-shooting-sister-video.

15 Kendi Anderson, "Woman Shoots Up Hixson Neighborhood," *Chattanooga Times Free Press*, December 28, 2014, https://www.timesfreepress.com/news/local/story/2014/dec/28/womshoots-hixsneighborhood/280032/.

16 Aamna Mohdin, Peter Walker and Nazia Parveen, "No 10's Race Report Widely Condemned as 'Divisive,'" *Guardian*, March 31, 2021.

17 Government of the United Kingdom, *Commission on Race and Ethnic Disparities: The Report*, March 2021, 232–34.

18 Sidanius and Pratto, *Social Dominance*, 228–29.

19 Brian A. Nosek et al., "Pervasiveness and Correlates of Implicit Attitudes and Stereotypes," *European Review of Social Psychology* 1, no. 52 (2007).

20 Sidanius and Pratto, *Social Dominance*, 229–31.

21 Nosek, Banaji, and Greenwald, "Harvesting Implicit Group Attitudes and Beliefs," 101–15.

22 Sidanius and Pratto, *Social Dominance*, 229–31.

23 Edward Selby et al., "Self-Sabotage: The Enemy Within," *Psychology Today*, September 2, 2011, https://www.psychologytoday.com/ca/articles/201109/self-sabotage-the-enemy-within.

24 "Javier Espinoza: Turning Pain into Power," in "5 Brave Personal Stories of Domestic Abuse," *TEDBlog*, January 25,2013, https://blog.ted.com/5-brave-personal-stories-of-domestic-abuse/.

25 Sidanius and Pratto, *Social Dominance*, 59.

26 Sidanius and Pratto, 248–49.

27 Sidanius and Pratto.

28 Sidanius and Pratto, 256–62.

29 Rob Kunzia, "Racism in Schools: Unintentional but No Less Damaging," *Pacific Standard*, April 8, 2009, https://psmag.com/education/racism-in-schools-unintentional-3821.

30 Ontario Human Rights Commission, "Human Rights Settlement Reached with Ministry of Education," 2005; Travis Riddle and Stacey Sinclair, "Racial Disparities in School-Based Disciplinary Actions Are Associated with County-Level Rates of Racial Bias," *Proceedings of the National Academy of Sciences of the United States of America*, April 2, 2019; Kuba Shand-Baptiste, "UK Schools Have Targeted Black Children for Generations–the Education System is Overdue for a Reckoning," *Independent*, January 20, 2020.

31 Sidanius and Pratto, *Social Dominance*, 103–25.

32 Max Weber, *The Protestant Ethic and the Spirit of Capitalism* (Taylor & Francis e-Library, 2005).

33 Weber, 106; Lydia Saad, "Black-White Educational Opportunities Widely Seen as Equal," *Gallup*, July 2, 2007; Frank Newport, "Little 'Obama Effect' on Views about Race Relations," *Gallup*, October 29, 2009.

34 Jeff Jones and Lydia Saad, "Gallup News Service, June Wave 2," *Politico*, June 19-30, 2019, https://www.politico.com/f/?id=0000016b-f5b8-d507-ab6b-fffdc7dc0000.

35 Knowledge@Wharton, "To Increase Charitable Donations, Appeal to the Heart—Not the Head," Wharton School of the University of Pennsylvania, 2007, https://knowledge.wharton.upenn.edu/article/to-increase-charitable-donations-appeal-to-the-heart-not-the-head/.

36 Associated Press, "Obama Quits Church after Controversy," *USA Today*, June 1, 2008.

37 *Hannity & Colmes*, "Obama's Pastor: Rev. Jeremiah Wright" (transcript), *Fox News*, March 1, 2007, https://www.foxnews.com/story/obamas-pastor-rev-jeremiah-wright.

38 Vann R. Newark II, "The Language of White Supremacy," *The Atlantic*, October 6, 2017.

39 Jon Greenberg, "Black Lives Matter Protests and the Capitol Assault: Comparing the Police Response," PolitiFact: The Poynter Institute, January 8, 2021.

40 Amy Goodman, "'We Never Made It to the Polls': Police in North Carolina Pepper-Spray Voting March, Arresting Eight," *Democracy Now*, November 2, 2020.

41 Associated Press, "FBI Says It Warned about Possible Violence ahead of U.S. Capitol Riot," January 13, 2021.

42 Geoff Ward, "Living Histories of White Supremacist Policing: Towards Transformative Justice," *Du Bois Review: Social Science Research on Race* 15, no. 1 (July 2018): 167–84.

SIX: Power Part 2: This Time It's Personal

1 The Process Work approach to rank and power comes from the work of Arnold Mindell, author of *The Deep Democracy of Open Forums* (Hampton Roads, 2002) and *Sitting in the Fire* (Deep Democracy Exchange, 2014).

2 Robin DiAngelo, "White Fragility: Why It's So Hard to Talk to White People about Racism," *The Good Men Project*, April 9, 2015.

3 Kate Allen, "Suspension Lengthened for Coach Who Opposed Slur," *Toronto Star*, December 17, 2010, https://www.thestar.com/news/gta/2010/12/17/suspension_lengthened_for_coach_who_opposed_slur.html.

4 Clea Skopeliti, "BLM Supporter Speaks Out after Carrying Counter-protester to Safety," *Guardian*, June 14, 2020.

5 Heterodox Academy, "Fighting Racism without Shaming with Irshad Manji," YouTube video, August 21, 2020.

6 Rhonda V. Magee, *The Inner Work of Racial Justice: Healing Ourselves and Transforming Our Communities through Mindfulness* (New York: Tarcher-Perigee, 2019), 32.

7 Goleman, Boyatzis, and McKee, *Primal Leadership*, 51–52.

8 Julie Diamond, interview with Shakil Choudhury, July 2, 2014.

9 Diamond, interview with Choudhury, July 2, 2014.

10 Jon Henley, "I Feel Terribly Guilty and Very Much Afraid," *Guardian*, November 4, 2004, https://www.theguardian.com/guardianweekly/story/0,,1349973,00.html.

SEVEN: Deep Diversity: Bringing It All Together

1 Asubpeeschoseewagong First Nation, www.grassynarrows.ca.
2 "Mercury Poisoning Effects Continue at Grassy Narrows," *CBC News*, June 4, 2012, https://www.cbc.ca/news/canada/mercury-poisoning-effects-continue-at-grassy-narrows-1.1132578.
3 Jody Porter, "'Longest Running' First Nations Blockade Continues," *CBC News*, December 3, 2012, https://www.cbc.ca/news/canada/thunder-bay/longest-running-first-nations-blockade-continues-1.1161095; Mike Aiken, "Grassy Members Remain Vigilant about Logging," *KenoraOnline*, August 22, 2020, https://kenoraonline.com/local/grassy-members-remain-vigilant-about-logging.
4 Ian C. Fisher et al., "The Relationship between Meaning in Life and Post-Traumatic Stress Symptoms in US Military Personnel: A Meta-Analysis," *Journal of Affective Disorders* 277, December 1, 2020.
5 Steven M. Southwick and Dennis S. Charney, "Ready for Anything," *Scientific American Mind* (July/August 2013): 32–41.
6 Southwick and Charney.
7 The details of Francine's story and the Oka crisis are well documented in Loreen Pindera, "A Sister's Grief Bridges a Cultural Divide: Revisiting the Oka Standoff," *CBC News*, July 8, 2010, https://www.cbc.ca/news/canada/a-sister-s-grief-bridges-a-cultural-divide-1.971486; and Loreen Pindera, "Bringing Down the Barricades," originally aired on CBC Radio's *C'est la Vie* with Bernard St-Laurent in June 2010, https://www.cbc.ca/player/play/2266962406.
8 Ingrid Peritz, "Sister of Slain Officer at Oka Makes Peace with Mohawks," *Globe and Mail*, July 4, 2010, https://www.theglobeandmail.com/news/national/sister-of-slain-officer-at-oka-makes-peace-with-mohawks/article1386578/.
9 Peritz.
10 Facts and story confirmed through personal communication with Francine Lemay, February 8, 2021.
11 Pindera, "A Sister's Grief."
12 Pindera.
13 Pindera.
14 Pindera, "Bringing Down the Barricades."
15 Pindera, "A Sister's Grief."
16 Pindera, "Bringing Down the Barricades."
17 Lawrence Martin, "A Democracy That's More like Putin's Than Lincoln's: Is That What Republicans Want?," *Globe and Mail*, June 2, 2021.
18 Levitsky and Ziblatt, *How Democracies Die*.
19 Othering & Belonging Institute, "Bridging —Towards A Society Built on Belonging: Animated Video + Curriculum."
20 Krista Tippett, "Derek Black and Matthew Stevenson: Befriending Radical Disagreement," *On Being*, May 17, 2018.

Index

PRAISE FOR **DEEP DIVERSITY**

"Everyone working on race issues should read this book. Even when you don't agree, you will be provoked to think harder about the enormity of our challenge, and how to generate the emotional, as well as intellectual, fortitude to meet that challenge."
RINKU SEN, author of *The Accidental American*

"*Deep Diversity* provides a panoramic view of our social landscape and a deep dive into issues of implicit bias, personal and systemic power dynamics, and the potential for healing and racial justice."
JOSEPH GOLDSTEIN, author of *Mindfulness: A Practical Guide to Awakening*

"*Deep Diversity* offers an important analysis to help us achieve the genuine reconciliation that we must achieve between Canadians and Indigenous Peoples in order to move forward."
ARTHUR MANUEL, Neskonlith, Secwepemc Nation, coauthor of *Unsettling Canada: A National Wake-Up Call*

"Choudhury provides an open, honest, and plainspoken view of diversity issues. [His] approach provides a method by which both sides of any disagreement can be empowered to join the conversation and actually 'want' change." ***CHOICE CONNECT***

"*Deep Diversity*... coherently presents scientific evidence, a systems thinking analysis of structural racism as well as mindfulness and self-care as much-needed and interconnected foundations for authentic personal and social change from within."
GERY PAREDES VÁSQUEZ, race and gender equity director, YWCA Madison

"*Deep Diversity* is a groundbreaking book taking a giant step towards overcoming pervasive racism in our society."
JUDY REBICK, author of *Heroes in My Head* and *Ten Thousand Roses*

"Scholarly, inspiring, and full of hope, this is a book full of wisdom of how to create a better world." **PAUL GILBERT,** author of *The Compassionate Mind*